ENGINEERING TOOLS, TECHNIQUES AND TABLES

MACHINE TOOLS: DESIGN, RELIABILITY AND SAFETY

ENGINEERING TOOLS, TECHNIQUES AND TABLES

Additional books in this series can be found on Nova's website under the Series tab.

Additional E-books in this series can be found on Nova's website under the E-book tab.

ENGINEERING TOOLS, TECHNIQUES AND TABLES

MACHINE TOOLS: DESIGN, RELIABILITY AND SAFETY

SCOTT P. ANDERSON
EDITOR

Nova Science Publishers, Inc.
New York

Copyright © 2011 by Nova Science Publishers, Inc.

All rights reserved. No part of this book may be reproduced, stored in a retrieval system or transmitted in any form or by any means: electronic, electrostatic, magnetic, tape, mechanical photocopying, recording or otherwise without the written permission of the Publisher.

For permission to use material from this book please contact us:
Telephone 631-231-7269; Fax 631-231-8175
Web Site: http://www.novapublishers.com

NOTICE TO THE READER

The Publisher has taken reasonable care in the preparation of this book, but makes no expressed or implied warranty of any kind and assumes no responsibility for any errors or omissions. No liability is assumed for incidental or consequential damages in connection with or arising out of information contained in this book. The Publisher shall not be liable for any special, consequential, or exemplary damages resulting, in whole or in part, from the readers' use of, or reliance upon, this material. Any parts of this book based on government reports are so indicated and copyright is claimed for those parts to the extent applicable to compilations of such works.

Independent verification should be sought for any data, advice or recommendations contained in this book. In addition, no responsibility is assumed by the publisher for any injury and/or damage to persons or property arising from any methods, products, instructions, ideas or otherwise contained in this publication.

This publication is designed to provide accurate and authoritative information with regard to the subject matter covered herein. It is sold with the clear understanding that the Publisher is not engaged in rendering legal or any other professional services. If legal or any other expert assistance is required, the services of a competent person should be sought. FROM A DECLARATION OF PARTICIPANTS JOINTLY ADOPTED BY A COMMITTEE OF THE AMERICAN BAR ASSOCIATION AND A COMMITTEE OF PUBLISHERS.

Additional color graphics may be available in the e-book version of this book.

Library of Congress Cataloging-in-Publication Data

Machine tools : design, reliability and safety / editor, Scott P. Anderson.
   p. cm.
 Includes index.
 ISBN 978-1-61209-144-0 (hardcover)
 1. Machine-tools. 2. Machine-tools--Design and construction. I. Anderson, Scott P.
 TJ1185.M223 2010
 621.9'02--dc22
                       2010047092

Published by Nova Science Publishers, Inc. † New York

CONTENTS

Preface		vii
Chapter 1	Manufacturing Process Planning Based on Reliability Analysis of Machines and Tools *Carmen Elena Patino Rodriguez and Gilberto Francisco Martha de Souza*	1
Chapter 2	Dynamic Study of a Centerless Grinding Machine through Advanced Simulation Tools *Iker Garitaonandia, Joseba Albizuri, M. Helena Fernandes, Jesús Mª Hernández and Itxaso Olabarrieta*	39
Chapter 3	Light Machine Tools for Productive Machining *J. Zulaika, F. J. Campa and L. N. Lopez Lacalle*	81
Chapter 4	Improving Machine Tool Performance through Structural and Process Dynamics Modeling *Tony L. Schmitz, Jaydeep Karandikar, Raul Zapata, Uttara Kumar, and Mathew Johnson*	117
Chapter 5	Optimum Design of a Redundantly Actuated Parallel Manipulator Based on Kinematics and Dynamics *Jun Wu, Jinsong Wang, Liping Wang and Tiemin Li*	153
Chapter 6	Site Characterization Model Using Machine Learning *Sarat Das, Pijush Samui and D. P. Kothari*	175
Chapter 7	Does Miniaturization of NC Machine-Tools Work? *Samir Mekid*	187
Chapter 8	Computer-Controlled Machine Tool with Automatic Truing Function of Wood-Stick Tool *Fusaomi Nagata, Takanori Mizobuchi, Keigo Watanabe, Tetsuo Hase, Zenku Haga and Masaaki Omoto*	193
Chapter 9	Fault Monitoring and Control of Mechanical Systems *S. N. Huang and K. K. Tan*	211
Index		233

PREFACE

In machine tools, the designed systems include many components, such as sensors, actuators, joints and motors. It is required that all these components work properly to ensure safety. This book examines fault monitoring and control schemes in machine systems, as well as detecting machines whenever a failure occurs and accommodating the failures as soon as possible. Also discussed are centerless grinding machines; improving machine tool performance through structural and process dynamics modeling and exploring the strength of the Japanese machine tool industry.

Chapter 1 - In many manufacturing industries, planning and scheduling are decision-making processes used as important roles in procurement and production. The planning and scheduling functions in a company rely on mathematical techniques and heuristic methods to allocate resources such as machine-tools, to execute the activities to comply with the production planning requirements. Objectives can take many different forms, such as minimizing the time to complete all activities, minimizing the number of activities or maximizing the process reliability. The manufacturing process must consider its dynamic as a mechanism to avoid non-conforming parts.

A manufacturing process is defined as a sequence of pre-established operations, aiming at the production of a specific part. The process reliability is depended on the operation sequence and their reliabilities. The operation reliability is a statistical relation between cutting tool, operator and machine-tool reliability. Machines often have to be reconfigured or cleaned between jobs. This is known as a changeover or setup. The length of the setup depends on the job just completed and on the one about to be started.

This chapter presents machining process planning based on reliability analysis of machines and tools. It is possible to determine the running time for each tool involved in the process by obtaining the operations sequence for the machining procedure. The cutting tool life can be modeled with a probability function representing the chance that a critical wear magnitude is achieved in a given time period. The machine-tool reliability can be modeled according to an exponential distribution, which parameter is dependent on the electrical motor, spindle and bearings failure rates. Aiming at keeping the manufacturing process reliability higher than a minimum value, defined in the process planning, an algorithm to define the cutting tool change period is presented.

Chapter 2 - Centerless grinding operations present some characteristic features, which make the process especially prone to suffering dynamic instabilities, leading to chatter vibrations. This kind of self-excited vibrations are very pernicious, as they not only limit the quality of the machined parts, but also the lifetime of the machine and the tool.

The design of control systems adapted to reduce vibrations requires the availability of accurate models capable of evaluating the effectiveness of different control alternatives prior to their practical implementation, so an optimization process can be tackled via simulations.

In this chapter a design procedure is presented to obtain reliable low-order models of the dynamic response of a centerless grinding machine. For this purpose, an approach based on the combined use of both numerical and experimental techniques was selected. Initially, the finite element (FE) model of the machine was updated taking as reference data obtained from an experimental modal analysis (EMA). This updated model showed to accurately characterize the vibration modes excited in the machine structure under chatter conditions. Nevertheless, the major drawback is that it has a large number of degrees of freedom, implying computationally expensive calculations. Thus, in a second step, the updated FE model was reduced to obtain a low order state space model covering the dynamic characteristics of the machine in the frequency range of interest.

The reduced model was integrated in the chatter loop of the centerless grinding process, and several simulations were performed both in the frequency and time domains. Different machine responses were estimated in these simulations and the results were compared to those obtained in machining tests, achieving good agreement between the theoretical and experimental results and, hence, validating the reduced model.

Chapter 3 - This chapter presents an integrated 'machine and process' approach that is based on stability lobe diagrams for designing large-volume milling machines, bearing in mind their productivity, reliability and accuracy as well as their eco-efficiency. In fact, eco-efficiency of machining processes is an issue of increasing concern among both machine tool builders and manufacturers. Thus, this chapter introduces a global modeling of the dynamics of milling machines and of milling processes with a goal of supporting engineers in the design of machine tools that result in optimal machining productivity at minimized environmental impacts and costs. This is a result of reducing the machines' material content, and consequently resulting in lowered energy consumption. This approach has been applied to the design of an actual milling machine, upon which machining tests have been conducted, that have shown an increase of 100% in productivity while consuming 15% less energy. This result is due to a weight reduction of over 20% in structural components, thus integrating highly productive machining processes and eco-efficient milling machines in one unique system.

Chapter 4 - There are many factors that influence the performance of computer-numerically controlled machining centers. These include tool wear, positioning errors of the tool relative to the part, spindle error motions, fixturing concerns, programming challenges, and the machining process dynamics. In this study, the limitations imposed by the process dynamics are considered. Algorithms used to predict the tool point frequency response function and, subsequently, the stability and surface location error (due to forced vibrations) are described. This information is presented graphically in the form of the milling "super diagram", which also includes the effect of tool wear and incorporates uncertainty in the form of user-defined safety margins. Given this information at the process planning stage, the programmer can select optimized operating parameters that increase the likelihood of first part correct production and reduce the probability of damage to the tool, spindle, and/or part due to excessive forces and deflections.

Chapter 5 - Parallel manipulators have attracted much attention in both industry and academia because of their conceptual potentials in high motion dynamics and accuracy combined with high structural rigidity due to their closed kinematic loops. However, there is a gap between the expectation and practical application of parallel manipulators in the machine tool/robot sectors. One of the reasons is that their potentially desirable high dynamics can not be realized since the dynamic characteristics are not considered in the kinematic design phase. The dynamic characteristics can be considered in the model-based control after the prototype is built. However, once the prototype is fabricated, the improvement of dynamic characteristics is limited even if a model-based control is used. If the dynamic characteristics can also be involved in the process of the kinematic design before the prototype is built, the dynamic performance would be improved more. It is helpful to realize the high motion dynamics of parallel manipulators.

This chapter presents a new method for the optimum design of parallel manipulators by taking both the kinematic and dynamic characteristics into account. The optimum design of a 3-DOF redundant parallel manipulator with actuation redundancy is investigated to demonstrate the method. The dynamic model is derived and a dynamic manipulability index is proposed. Based on the results of kinematic optimal design, the manipulator is re-designed by considering the dynamic performance. The kinematic performance may be debased, but the dynamic performance is improved. By using the method proposed in this chapter, the designer can obtain the optimum result with respect to both kinematic performance indices and dynamic performance in dices. Since the dynamic performance is considered in the process of optimum design by using the method proposed in this chapter, it is expected to realize the high dynamics of parallel manipulators.

Chapter 6 - This chapter describes two machine learning techniques for developing site characterization model. The ultimate goal of site characterization is to predict the in-situ soil properties at any half-space point for a site based on limited number of tests and data. In three dimensional analysis, the function $N = N(X, Y, Z)$ where X, Y and Z are the coordinates of a point corresponds to Standard Penetration Test(SPT) value(N), is to be approximated with which N value at any half space point in site can be determined. The site is located in the alluvial Gangetic plane (Sahajanpur of Uttar Pradesh, India). The input of machine leaning techniques is X, Y and Z. The output of machine learning techniques is N. The first machine learning technique uses generalized regression neural network (GRNN) that are trained with suitable spread(s) to predict N value. The second machine learning technique uses Least Square Support Vector Machine (LSSVM), is a statistical learning theory which adopts a least squares linear system as a loss function instead of the quadratic program in original support vector machine (SVM). Here, LSSVM has been used as a regression technique. The developed LSSVM model has been used to compute error bar of the predicted data. An equation has been also developed for the prediction of N value based on the developed LSSVM model. A comparative study between the two developed machine learning techniques has been presented in this chapter. This chapter shows that the developed LSSVM model is better than GRNN.

Chapter 7 - An immediate need was expressed recently and has triggered the process of miniaturization of machines extensively over the last decade. The requirement was about micro-mesoscale components with high to ultra high precision dimensions and surface finish but achieved at low cost according to comparison studies between standard scale machines and desktop machines.

The interest is for the development of hybrid subtractive/additive desktop meso/micro NC Machine-tool capable to secure a continuum of manufacturing capabilities and constitute a bridge between well developed technologies at both scales. Although the concept of miniaturization is reasonably justifiable, the challenge is whether in practice current miniaturized machines can achieve the expected requirements?

The next concern is whether the microfabrication techniques at meso-scale level, such as lithography, LIGA and their variations to produce MEMS type components and others are capable to improve and deliver better components in terms of complexity of shapes and dimensional precision.

Chapter 8 - In this chapter, a computer-controlled machine tool with an automatic truing function of a wood-stick tool is described for the long-time lapping process of an LED lens cavity. The authors have experimentally found that a thin wood stick with a ball-end tip is very suitable for the lapping tool of an LED lens cavity. When the lapping is conducted, a special oil including diamond lapping paste, whose grain size is about 3 μm, is poured into each concave area on the cavity. The wood material tends to compatibly fit both the metallic material of cavity and the diamond lapping paste due to the soft characteristics of wood compared with other conventional metallic abrasive tools. The serious problem in using a wood-stick tool is the abrasion of the tool tip. For example, an actual LED lens cavity has 180 concave areas so that the cavity can form small plastic LED lenses with mass production at a time. However, after about five concave areas are finished, the tip of the wood-stick tool is deformed from the initial ball shape as a result of the abrasion. Therefore, in order to realize a complete automatic finishing system, some truing function must be developed to systematically cope with the undesirable tool abrasion. The truing of a wood-stick tool means the reshaping to the initial contour, i.e. ball-end tip. In the chapter, a novel and simple automatic truing method by using cutter location data is proposed and its effectiveness and validity are evaluated through an experiment. The cutter location data are called the CL data, which can be produced from the main-processor of 3D CAD/CAM widely used in industrial fields. The proposed machine tool can easily carry out the truing of a wood-stick tool based on the generally-known CL data, which is another important feature of the authors' proposed machine tool.

Chapter 9 - In machine tools, the designed systems include many components, such as sensors, actuators, joints and motors. It is required to all these components should work properly to ensure safety. In this chapter, the authors will study fault monitoring and control scheme in the machine system, detecting machines whenever a failure occurs and accommodating the failures as soon as possible. In this scheme, the monitoring algorithm is first designed based on the system model. Under such circumstances, when a failure occurs, the fault-tolerant control scheme is activated to compensate the effects of the fault function. Case study is given to illustrate the performance of the designed monitoring and controller in a real-time machine.

In: Machine Tools: Design, Reliability and Safety
Editor: Scott P. Anderson, pp. 1-38

ISBN: 978-1-61209-144-0
© 2011 Nova Science Publishers, Inc.

Chapter 1

MANUFACTURING PROCESS PLANNING BASED ON RELIABILITY ANALYSIS OF MACHINES AND TOOLS

Carmen Elena Patino Rodriguez[1,2*]
and Gilberto Francisco Martha de Souza[2]

[1] Industrial Engineering, University of Antioquia,
Medellín, Calle 67 #53 – 108 – Ciudad Universitaria,
Medellín, Antioquia, Colombia
[2] Polytechnic School, University of São Paulo,
São Paulo, Av. Prof. Mello Moraes, 2231, Cidade Universitária,
São Paulo, SP, Brazil

ABSTRACT

In many manufacturing industries, planning and scheduling are decision-making processes used as important roles in procurement and production. The planning and scheduling functions in a company rely on mathematical techniques and heuristic methods to allocate resources such as machine-tools, to execute the activities to comply with the production planning requirements. Objectives can take many different forms, such as minimizing the time to complete all activities, minimizing the number of activities or maximizing the process reliability. The manufacturing process must consider its dynamic as a mechanism to avoid non-conforming parts.

A manufacturing process is defined as a sequence of pre-established operations, aiming at the production of a specific part. The process reliability is depended on the operation sequence and their reliabilities. The operation reliability is a statistical relation between cutting tool, operator and machine-tool reliability. Machines often have to be reconfigured or cleaned between jobs. This is known as a changeover or setup. The length of the setup depends on the job just completed and on the one about to be started.

This chapter presents machining process planning based on reliability analysis of machines and tools. It is possible to determine the running time for each tool involved in the process by obtaining the operations sequence for the machining procedure. The

---

[*] Corresponding author, e-mail: cpatino@udea.edu.co, phone: 55-11-3091-9847.

cutting tool life can be modeled with a probability function representing the chance that a critical wear magnitude is achieved in a given time period. The machine-tool reliability can be modeled according to an exponential distribution, which parameter is dependent on the electrical motor, spindle and bearings failure rates. Aiming at keeping the manufacturing process reliability higher than a minimum value, defined in the process planning, an algorithm to define the cutting tool change period is presented.

1. INTRODUCTION

Reliability is defined as the probability that a given product will successfully perform a required function without failure, under specified environmental conditions, for a specified period of time.

Deficiencies in design and manufacturing planning affect all items produced and are progressively more expensive to correct as development proceeds. It is therefore essential that design and manufacturing disciplines be used to minimize the possibility of failure and to allow design deficiencies to be detected and corrected as early as possible. The reliability concepts are usually applied in the product design process but its use in the manufacturing process planning is not so common.

In 2000, Savsar presented a stochastic model to determine the performance of a Flexible Manufacturing Cell under random operational conditions, including random failures of cell components (machine tool and robot) in addition to random processing times, random machine loading and unloading times, and random pallet transfer times [1].

A methodology based on Petri nets was used for identifying the failure sequences and assessing the probability of their occurrence in the manufacturing system. The method employs Petri net modeling and reachability trees constructed based on the Petri nets. The methodology is demonstrated on an example of an automated machining and assembly system [2].

Manufacturing processes planning is an activity executed by any mechanical industry in order to define the processes and the tools that will be used to manufacture specific parts and to assembly a specific system. Usually that activity involves the theoretical and experimental evaluation of machine tools' accuracy and resolution, tools performance, and once the manufacturing processes is defined, a try out is executed, aiming the definition of the processes capability. The basic goal of processes planning is to define a sequence of manufacturing activities to produce a mechanical part according to design dimensional and geometrical tolerances.

Considering any manufacturing process as a system, composed of machine tools, tools, and tools operational conditions, its reliability can be evaluated using the traditional reliability concepts, aiming the definition of a manufacturing process failure probability, taking in view the occurrence of failures in the components of the systems mentioned above.

2. PROCESS PLANNING IN MANUFACTURING SYSTEMS

The manufacturing process planning is a complicated and combined problem, therefore it is necessary to divide the tasks into hierarchical levels, [3]. In this manner, the process plan is defined successively, step by step, up to down. The 'up to down' term means that the planning process progresses from the complex tasks to the simpler and the process plan becomes more detailed and concrete.

In technical literature there is no unified standpoint about the number and the tasks of these planning levels. Groover [4] divides the whole process planning as follows: (1) preliminary process planning, (2) planning the sequence of operations, (3) operations planning, (4) operation elements planning.

The preliminary process planning is the highest level of manufacturing process planning, the strategy of the manufacturing. According to those authors, the tasks of the process planning are as follows:

- Collecting the technological data for the process planning of the blank manufacturing, the part manufacturing and the assembly; rationalizing of manufacturing process; preparing of manufacturability and assembly of the correct part, assembly and blank design documentation.
- Determining the strategy of process planning which means the selection of manufacturing systems and actual manufacturing variant.
- Analyzing the manufacturing tasks, estimation of manufacturing cost and time data.

Based on the above mentioned definitions, it is possible to observe that some information is not considered by those authors, such as:

- The manufacturing operation sequence must considerer the time to produce a given lot of parts. The operational time is important once the machine tool operational condition and the tool wear are time-dependent and can affect the manufactured part precision.
- The tool and machine tool reliability deal with modeling the time dependent performance of those components and can allow the prediction of possible failures in the manufacturing processes in advance. Those failures are related to the loss of geometrical and dimensional precision of the manufactured parts. The use of reliability concepts can anticipate problems in manufacturing and together with the traditional quality control procedure (usually based on Quality Control Charts), would improve the early detection of possible manufacturing process loss of quality.
- The use of reliability concepts can also be used to detect possible weaknesses in the manufacturing process associated with potential failures in machine tool that could significantly affect the process reliability.

Based on the previous analysis, it is important to add reliability concepts to manufacturing process planning for the prediction of loss of accuracy in the produced parts associated to time-dependent machine and tool degradation phenomena. In order to correct add reliability concepts in manufacturing planning activities, the next section will discuss manufacturing planning functions and reliability concepts.

2.1. Process Planning Function

The functions of the process planning activities are related to the definition of the manufacturing and assembly sequence of a given product design. In order to achieve that goal, the process planning function is dependent on the available resources in the manufacture plant (including machine tools, cutting of forming tool characteristics and labor skill) and on the dimensional and/or geometrical tolerance of the parts.

Nevertheless, the general function of the manufacturing planning is to define the optimized sequence of process aiming at producing and assembling a given set of parts to compose a product. The manufacturing process planning function also includes the definition of the basic parameters of each manufacturing process used for each part, including tool selection, machine tool operational conditions, quality controls methods and process capability analysis. Also, in a broad way, the manufacturing planning process can also include the manufacturing plant capacity estimate and the development of manufacturing patterns.

Rozenfeld [5] proposed to divide the manufacturing process planning functions in two major tasks: Macro Planning and Detailed Planning as shown in Figure 1. In the first task, the manufacturing sequence is defined based on plant restrictions and in the second task, each manufacturing process is detailed, including machine tool, cutting or forming tool selection and estimative of manufacturing time.

Rezende [6] proposes another view of the manufacturing process planning for a given part. The method can be considered complementary to the method proposed by Rozenfeld [5]. The first step corresponds to a general manufacturing planning for the part where the manufacturing steps for part production are defined according to design specifications. The second step corresponds to a detailed manufacturing planning where each manufacturing step is specified; including machine tool and tools, selections and geometrical and dimensional tolerance studies are executed in order to define jigs and fixtures selections. In Table 1, the method basic steps are presented.

It is important to notice that those methods do not consider the possible aging effects associated with the machine tools that could affect their accuracy and consequently the parts dimensional and geometrical tolerances. For those methods, the machine tools have the same performance during its operational life.

Furthermore, during cutting of forming tool selection, those methods consider the possible aging effects presented by the tools, but in a deterministic manner. The manufacturing planner frequently uses empirical relations (such as the Taylor Eq. for cutting tool) to estimate the tool life aimed at controlling the tool wear.

The reliability concepts can be used to predict the time-dependent aging effects acting on tools and machine tools aimed at predicting the frequency of failure of the manufacturing process associated with equipment failure. The reliability concepts can also be used to estimate tool wear aimed at defining tool change time during manufacturing planning.

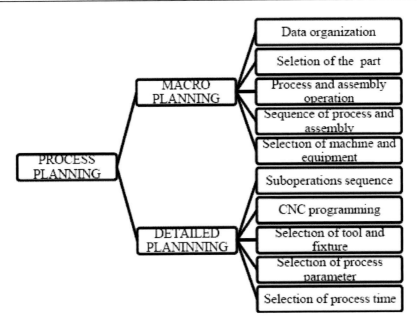

Figure 1. Manufacturing Process Steps [5].

Table 1. Manufacturing Process Planning Steps [6]

Step	Sub-Step	Characteristics under Analysis	
General Process Planning	Part design analysis	Part geometry	
		Basic surfaces definitions	
		Material and heat treatment	
		Part dimensions	
	Preliminary machining methods	Surface geometry	
		Surface dimensions	
		Surface finishing requirements	
		Part weight and size	
		Material and heat treatment	
		Production volume	
	Definition of manufacturing process steps	Operations	Rough
			Half-finish
			Finish
	Operations grouping	Tools and Machine Tools capability	
		Machine tools availability	
		Set up time and operation time	
		Part size and weight	
		Production volume	
	Heat treatment grouping	Heat treatment	Normalizing
			Annealing
			Quenching
			Temper
			Others
	Auxiliary manufacturing steps	Part characteristics	
		Finishing	
		Inspection	
		Cleaning	

Table 1. (Continued)

Step	Sub-Step	Characteristics under Analysis	
Detailed process planning	Machine tools selection	Machine precision x part required precision	
		Machine working area	
		Machine power	
		Machine capacity x Part production volume	
		Machine availability	
	Machine tool selection	Tool	Geometry
			Material
		Machining process	
		Machine tool selection	
		Part surface finishing precision	
	Jigs and fixtures selection	Set up time	
		Operational costs	
Detailed process planning	Reference surfaces selection	Reference type	Design
			Manufacturing
			Measuring
			Assembly
		Design dimensions and tolerances	
		Manufacturing dimensions and tolerances	
	Machining dimensions and tolerances	Prior operation surface finishing	
		Machining conditions	
	Cutting conditions selection	Cutting depth	Machine tool power
			Cutting tool characteristics
			Part stiffness
			Fixture
		Feed rate	Surface finishing
		Cutting speed	Production volume
			Production costs
	Reference time	Machining time	
		Set up time	
	Process planning documents	Process sequence	
		Detailed operation plan	

3. RELIABILITY CONCEPTS

Reliability has many connotations. In general, reliability is defined as the probability that a given product will successfully perform a required function without failure, under specified environmental conditions, for a specified period of time [7].

A reliability measure is needed to answer questions such as: "How long will the equipment last without breakdown?"; "How long a warranty can be given for new equipment?". In any given equipment population, the life length is a random variable. When questions must be answered regarding the behavior of such random variables, it is necessary to use probability, probability distributions, averages and measures of variability.

It must be realized, while defining the reliability of a piece of equipment, that the function the equipment is expected to perform must be clearly specified along with a definition of what constitutes a failure [8].

Furthermore, the conditions under which the piece of equipment is expected to perform the required function must also be clearly specified. Reliability under one set of operating conditions may be different from that under another.

Finally, reliability must be expressed as a function of time. At any specified time, a certain proportion of the equipment population will continue to successfully perform the required function without failure. Reliability can represent this proportion of the population that survives beyond the specified time [7].

Probably the single most used parameter to characterize reliability is the mean time to failure (or MTTF). It is just the expected or mean value of the failure time, expressed according to Eq. (1):

$$MTTF = \int_0^\infty R(t)dt \qquad (1)$$

where:

$R(t)$ reliability at time t

T time period, usually expressed in hours for pieces of equipment in power plants [h]

Random failures (represented by the exponential probability function) constitute the most widely used model for describing reliability phenomena. They are defined by the assumption that the rate of failure of a system is independent of its age and other characteristics of its operating history. In that case, the use of mean time to failure to describe reliability can be acceptable once the exponential distribution parameter, the failure rate, is directly associated with MTTF [9].

When the phenomena of early failures, aging effects, or both, are presented, the reliability of a device or system becomes a strong function of its age.

The Weibull probability distribution is one of the most widely used distributions in reliability calculations involving time-related failures. Through the appropriate choice of parameters, a variety of failure rate behaviors can be modeled, including constant failure rate, in addition to failure rates modeling both wear-in and wear-out phenomena [9].

The two-parameter Weibull distribution, typically used to model wear-out or fatigue failures is represented by the following Eq.:

$$R(t) = e^{-\left(\frac{t}{\eta}\right)^\beta} \qquad (2)$$

where:

$R(t)$ reliability at time t

t time period, usually expressed in hours for pieces of equipment in power plants [h]

β Weibull distribution shape parameter

η Weibull distribution characteristic life [h]

The life time distribution of a piece of equipment is the basic information from which all measures of reliability are evaluated. It is the distribution of the length of life of all items in the population of a piece of equipment. The distribution can be estimated from a set of sample life data taken from the population. Such data can be generated by testing a sample in the laboratory, or, as usually done for large pieces of equipment such as a machine tool, observing it in actual field use [10].

Failure rate is an important function in reliability analysis since it represents the changes in the probability of failure over the lifetime of equipment or component or even cutting or forming tool. Failure rate at any given time is the proportion of items that will fail in the next unit of time, out of those units that have survived up to that time.

The failure rate can increase, decrease or remain constant over time depending on the equipment's nature. Failure rate can be evaluated from the knowledge of reliability of the life (or reliability) distribution. How the failure rate changes over time gives an insight into the failure mechanisms of equipment and is used in studies for improving reliability. The failure rate curve that seems applicable to a wide variety of complex equipment is shown in Figure 2, known as bathtub curve due to its peculiar shape. Such a curve divides the life of equipment into three distinct regions, [7].

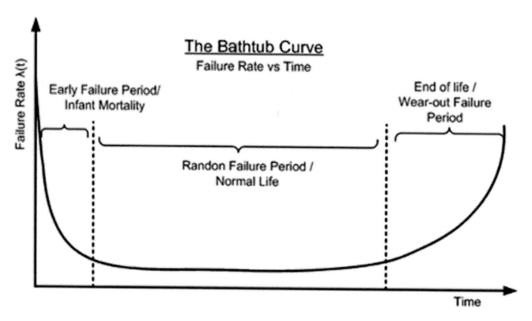

Figure 2. Failure Rate Curve – Bathtub Curve.

The, so-called burn-in early failure region exhibits decreasing failure rate. That decrease is due to defective units in the population, caused by poor material or failures in manufacturing processes, failing early and being repaired or removed from the population. The chance failure region presents an almost constant failure rate. Here, failures occur not because of inherent defects in the units but because of accidental occurrence of loads in excess of the design strength [9].

The constant failure rate approximation is often quite adequate even though a system or some of its components may exhibit moderate early-failures or aging effects. The magnitude of early failures is limited by strictly quality control in manufacturing and aging effects can be sharply limited by careful predictive or preventive maintenance.

Finally, the wear-out region is characterized by a complex aging phenomena, representing an increasing failure rate. Failures occur due to the development of cumulative damage mechanism such as wear, corrosion or fatigue. Knowledge of when wear-out begins helps in planning replacements and overhauls.

The failure rate $\lambda(t)$ at any time t is expressed according to Eq. (3):

$$\lambda(t) = \frac{f(t)}{R(t)} \tag{3}$$

where $f(t)$ is the probability density function associated with reliability distribution

The reliability of each equipment subsystem is calculated based on the failure data and the equipment reliability is simulated through the use of a block diagram. Reliability block diagram is frequently used to model, in a quantitatively way, the effect of item failures on system performance. It corresponds to the information flow arrangement among the items in the system. A block represents one or a collection of some basic parts of the system for which reliability data are available, [8].

Basically, there are two basic configurations of block diagrams: series or parallel systems. In a series system, the components are connected in such a manner that if one of the components fail the entire system fails, as shown in Figure 3a. In a parallel system, named as active redundancy, the system fails only when all of the components fail, as shown in Figure 3b. For that configuration, the redundant component is always alive. In this configuration, the system will perform its function if at least one of the components is working. In parallel standby systems, the redundant component is activated only after the main component has failed [9].

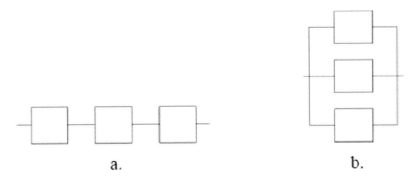

Figure 3. Reliability Block Diagram basic configurations. a) Series system b) Parallel system.

For a series system, considering that the component failures are independent probabilistic events, the system reliability in any time t is given as the product of component reliability and expressed according to Eq. (4):

$$R_s(t) = R_1(t).R_2(t)...R_k(t) \tag{4}$$

where $R_s(t)$ is system reliability at time t, $R_i(t)$ is i[th] component reliability at time t.

Table 2. Example of Process FMEA Worksheet Format

PROCESS FAILURE MODE AND EFFECTS ANALYSIS

PART NUMBER_____
REFERENCE DRAWING_____

DATE_____
SHEET___OF___

PROCESS FUNCTION/ REQUIREMENTS	POTENTIAL FAILURE MODE	POTENTIAL EFFECTS OF FAILURE	POTENTIAL CAUSES/ MECHANISM OF FAILURE	PROCESS CONTROL	SEVERITY	OCCURRENCE	DETECTION	RPN

Adapted from [9].

For an active parallel system, also considering the component failures as independent events, the system reliability is expressed according to Eq. (5):

$$R_s(t) = 1 - [1 - R_1(t)][1 - R2(t)]..[1 - R_k(t)] \qquad (5)$$

Complex pieces of equipment can be represented as a combination of series and parallel systems. The system must be modeled through a reliability block diagram and the reliability computed by evaluating reliability of subsystems in a bottom-up manner.

As a complement for the reliability numerical analysis, the manufacturing planner can develop a qualitative method to define the possible relations between the machine tool, tools, jigs and fixture failures failure modes and their effects on the manufacturing process, mainly on the product quality, the process reliability and human and environmental safety.

That analysis is performed based on the failure modes and affects analysis (FMEA) concepts and is commonly named Process FMEA (PFMEA). The PFMEA analysis can be developed according to the following steps:

i) For each process element (tool, machine tool, jigs and fixtures), determine the possible potential failure modes such as wear, lack of lubrication, incorrect assembly and others, where the occurrence can affect the performance of the manufacturing process.

ii) Describe the effects of those failure modes. For each failure mode identified, the PFMEA development team should determine what the ultimate effect will be on the manufacturing process. Examples of failure effects include: injury to the machine operators; inoperability of the machine tool affecting production planning; loss of precision in the part manufacturing (loss of part quality) and others.

iii) Identify the causes for each failure mode. A failure cause is defined as a design weakness that may result in a failure. The potential causes for each failure mode should be identified and documented.

iv) Enter the Probability factor. A numerical weight should be assigned to each cause that indicates how likely that cause is (probability of the cause occurrence). A common industry standard scale uses 1 to represent not likely and 10 to indicate inevitable.

v) Identify Current Controls (design or process). These are the mechanisms that prevent the cause of the failure mode from occurring or which detects the failure before it reaches the Customer. The team should identify testing, analysis, monitoring, and other techniques that can or have been used on the same or similar processes to detect failures. Each of these controls should be assessed to determine how well it is expected to identify or detect failure modes.

vi) Determine the likelihood of Detection. Detection is an assessment of the likelihood that the Current Process Controls will detect the Cause of the Failure Mode or the Failure Mode itself.

vii) Estimate Risk Priority Numbers (RPN). The Risk Priority Number is a mathematical product of the numerical Severity, Probability, and Detection ratings:

RPN = (Severity) x (Probability) x (Detection)

The RPN is used to prioritize items than require additional quality planning and maintenance requirements review.

viii) Determine Recommended Action(s) to address potential failures that have a high RPN. These actions could include specific inspection, testing or quality procedures; selection of different machines tools or tools; de-rating; limiting environmental stresses or operating range; monitoring mechanisms and performing preventative maintenance.

The FMEA analysis is executed with the use of a table such as the one presented in Table 2.

4. RELIABILITY FACTORS IN MACHINING MANUFACTURING PROCESS

This section of the chapter discusses the application of reliability concepts to model manufacturing process reliability.

Although those concepts can be applied to any manufacturing process, the main focus of this item is the machining process. The machine process is the most used manufacturing process in the mechanical industry, once parts that are manufactured by forming or casting usually are machined in order to control dimensional and geometrical tolerances and surface finishing.

Usually, the reliability of a manufacturing operation depends on three independent factors: operator, machine-tool and cutting tool. The reliability of each factor may be modeled by means of statistical distribution.

The following sections present some discussions regarding the failure modes and reliability models that can be used to define the machining process reliability.

4.1. Operator

People are affected by their environments in many ways. For example, social environment, heat or cold, light conditions and time pressure all have the potential of influencing human performance. One measure of the quality of work performance is the number of errors made by the people at work. Errors that are caused by human operators are referred to as 'human errors' [11].

Failure (or error) in this sense is usually defined as the failure to perform an act within the limits (of time or accuracy) required for safe system performance or the performance of a non-required act which interferes with system performance. The errors are classified as random and systematic errors.

Systematic errors are biases which lead to the situation where the mean of many separate measurements differs significantly from the actual value. Sources of systematic errors may be imperfect calibration of measurement instruments or incorrect use of the instrument by the operator. Systematic errors are detected applying statistical methods to the monitoring and control of a process and are corrected with frequent calibration of instruments and operator training.

Random errors are caused by unknown and unpredictable changes in the process. These changes may occur in the tools, in the environmental conditions or in complex man-machine interfaces. The random human errors introduce the possibility of manufacturing non-conforming parts, and this may become a function of physical or emotional fatigue of the operator. The effect of the random errors may be reduced by operator training or observation and averaging the outcomes. Furthermore, through the application of time and motion studies that consider the pauses at work and the division, workflows can be analyzed and synthesized with the objective of improving labor productivity.

The operator reliability may be defined experimentally based on the register of the number of errors that occurred during a specific period of observation.

4.2. Machine

The machine reliability depends on the machine architecture and design characteristics, including the degree of automation, the operational environment and also on the maintenance policy. However, as the breakdown of a single machine may result in the production of an entire workshop being halted and repairs may be more difficult and expensive when a breakdown occurs, the availability of the production line can be severely impacted by an unexpected machine failure. Wang et al. [12] show the main subsystems in a machine tool that present great frequency of failures which are the electric and electronic system, turret, CNC system, chuck and clamping fixture, power supply, servo unit. The most critical mechanical subsystems as for failure and reliability analysis are the turret and chuck (Figure 4).

Usually the reliability of each subsystem is different once it is dependent on the nature of the subsystem, such as mechanical or electrical/electronic. There are some subsystems with random failure such as the sensors and servo unit, while other subsystems present failures associated with damage accumulation processes for instance shaft, gear box, and bearings.

Researches were developed aimed at analyzing machine reliability used in mechanical parts manufacturing. Wang et al. [12] present a study about the failure modes and causes and define the weakest subsystems subsystem on CNC machines based on a probabilistic model developed through the study of failures in 80 CNC lathes, collected over a period of two years. This study established that the failure could be best described using the Lognormal distribution with mean 5.1758 and standard deviation 1.1370, the variable's natural logarithm, as shown in Figure 5.

Beginning-of-life equipment can present random failure (failure mode typical of electrical and electronic components), but after 100 hours of use without maintenance, the hazard failure is increased, as show in Figure 5, indicating the beginning aging stage where phenomena such as fatigue and wear of mechanical components are predominant which affects the machine performance.

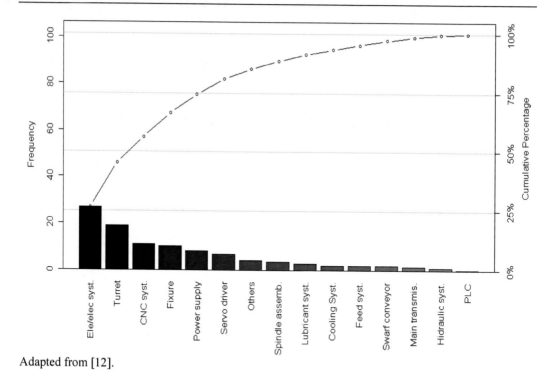

Adapted from [12].

Figure 4. Pareto chart of failure for CNC lathe.

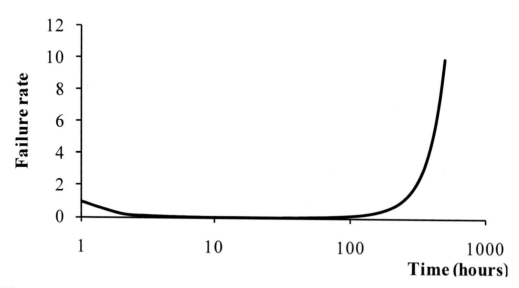

Figure 5. Failure rate for electromechanical machines – Lognormal Distribution (5.1758 , 1.1370).

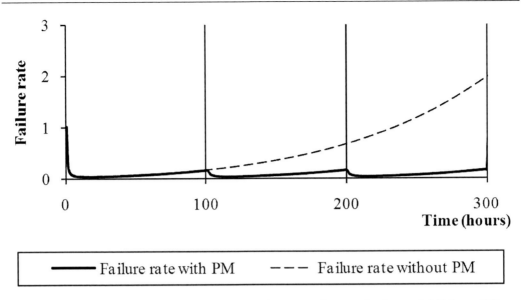

Figure 6. Behavior of failure rate with preventive maintenance. Intervention Interval: 100 hours [15].

The result of Wang et al. [12] conflict with other authors such as Mejabi and Black [13] and Freiheit and Hu [14] who established that the probability of failure in a manufacturing equipment can be modeled with the use of an exponential distribution. The exponential distribution would be used considering aspects such as: preventive maintenance, application of Failure Modes and Effect Analysis (FMEA) to improve the design of critical subsystems, and application of design methods based on "design life criterion" for the main mechanical components [15].

Preventive maintenance is a schedule of planned maintenance actions aimed at the prevention of unexpected breakdowns and failures. The primary goal of preventive maintenance is to prevent the failure of equipment before it actually occurs. It is designed to preserve and enhance equipment reliability by replacing worn components before they actually fail. Preventive maintenance activities include equipment checks, partial or complete overhauls at specified periods, oil changes, lubrication and so on. In addition, workers can record equipment deterioration so they can know the time to replace or repair worn parts before they cause system failure. The preventive maintenance program aims at preventing all equipment failure due to aging effects. The machine would ideally present random failures and its reliability can be modeled by an exponential distribution, as shown in Figure 6.

Nevertheless, the machine operational time without preventive maintenance must not exceed the beginning of the increased failure rate associated with the wear of power system components. Furthermore the mechanical main components as shaft and gear assembly are designed with infinite lifetime and high level of reliability. These facts can significantly delay the aging effects. Therefore, under these considerations it is possible to model the lifetime by an exponential distribution. The FMEA is an engineering technique used to define, identify, and eliminate known and/or potential failures, problems, and errors in the machines design enhancing their reliability.

4.3. Tool

During the machining process, the cutting tools are loaded with heavy forces resulting from the deformation process in chip formation and friction between the tool and work piece. The heat generated at the deformation and friction zones overheats the tool, the chip and partially, the work piece. All the contact surfaces are usually clean and chemically very active; therefore the cutting process is connected with complex physical-chemical processes. Tool wear, which occurs as the consequence of such processes, is reflected as progressive wearing of particles from the tool surface.

As for manufacturing process reliability analysis, the definition of cutting tool change time can be based on cutting tool reliability analysis. That analysis aims at keeping the tool reliability greater than a minimum target value that will reduce the chance of non-conforming parts manufacturing. The reliability of the cutting tool represents the probability that the tool wear is lower than a pre-defined value, in a given operational time.

The defects introduced in parts, as a function of tools failure, have increasing failure rate increase in time since the main tool failure mode is wear and the tool wear is accumulative damage mechanism. Therefore, the tool reliability must be represented by a probability distribution function that simulates increasing failure rates with time as the Normal distribution, Weibull distribution and Lognormal distribution. Some authors such as Freiheit and Hu [14], purpose the exponential distribution is appropriate to model the wear tool, which is not suitable to model failures associated with aging effects.

Some works aiming at modeling tool reliability are based on the methodology developed by Hitomi et al. [16], such as those developed by Wang et al. [17], El Wardany and Elbestawi [18], and Patino Rodriguez [15], where a reliability-dependent failure rate model is used to predict the reliability of a cutting tool subject to flank wear with Lognormal distribution.

It is assumed that the distribution of average flank wear (V_B) follows a Lognormal distribution. Whether tool wear fit another distribution, it is possible to follow this procedure and to find the tool reliability in terms of lifetime.

Let V_B be a random variable that represents the tool flank wear. The tool flank wear is a function (Ψ) of the cutting conditions and tool geometry. This function can be given by Eq. (6).

$$v_B = \psi(f, \upsilon, d, t, \gamma, r) \cdot \theta(\xi) \tag{6}$$

where f: feed rate (mm/rev), υ: cutting speed (m/min), d: depth of cut (mm), t: cutting time (min), γ: angle, r: radius, and θ: error.

By taking the logarithms of both sides of the Eq. 10, the following linear relation is obtained Eq. (7).

$$\begin{aligned}\ln(v_B) &= \ln[\psi(f,\upsilon,d,t,\gamma,r) \cdot \theta(\xi)] \\ \ln(v_B) &= \ln[\psi(f,\upsilon,d,t,\gamma,r)] + \ln[\theta(\xi)] \\ \ln(v_B) &= \ln[\psi(f,\upsilon,d,t,\gamma,r)] + \ln[\theta(\xi)]\end{aligned} \tag{7}$$

The median of flank wear (V_{B0}) and the variance (σ^2) represents the dispersion of the values for tool wear, the difference between the Taylor too-life equation value and actual wear value from experimental results. They are calculated from calculated from the relations Eqs. (8) and (9).

$$V_{B_0} = E[\ln(v_B)] = E[\ln(\psi(f,\upsilon,d,t,\gamma,r))] + E[\varepsilon] \qquad (8)$$

$$\begin{aligned}Var[\ln(v_B)] &= E[(\ln(v_B - V_{B_0})^2)] \Rightarrow \\ Var[\ln(v_B)] &= E[\varepsilon] \Rightarrow \\ Var[\ln(v_B)] &= \sigma^2\end{aligned} \qquad (9)$$

The relationship between tool flank wear V_B and the cutting condition assumed as the Taylor tool-life equations given by Eq. (10).

$$\hat{v}_B = C_0 \cdot f^{b_1} \cdot \upsilon^{b_2} \cdot d^{b_3} \cdot \gamma^{b_4} \cdot r^{b_5} \cdot t^{b_6} \qquad (10)$$

where C_0, b_1, b_2, b_3, b_4, b_5, b_6 are constants, which are determined from experimental results and $\hat{v}_B$ is standard wear, value estimated for the cutting conditions.

Assuming that the distribution of average flank wear V_B obeys a Lognormal distribution, the density functions of the flank wear is given by Eq. (11).

$$f(v_B) = \frac{1}{\sqrt{2\pi} \cdot \sigma \cdot v_B} e^{\left[-\frac{1}{2 \cdot \sigma^2}(\ln v_B - \ln(C_0 \cdot f^{b_1} \cdot \upsilon^{b_2} \cdot d^{b_3} \cdot \gamma^{b_4} \cdot r^{b_5} \cdot t^{b_6}))^2\right]} \qquad (11)$$

Suppose that a cutting tool begins to function at the time period of t=0, and that its failure occurs at t=T. Therefore, the probability that the tool fail at t, F(t), and tool reliability, R(t) are represented by Eqs. (12) and (13).

$$F(T) = P(t \le T) = \int_0^T f(t)dt \qquad (12)$$

$$\begin{aligned}R(T) &= 1 - F(T) \Rightarrow \\ R(T) &= 1 - \int_0^T f(t)dt\end{aligned} \qquad (13)$$

On the other hand, the probability that the tool life attain to end because the tool wear reached wear limit (V^*_B), is established by Eq. (14).

$$P(v_B \ge V_B^*) = 1 - \int_0^{v_B^*} f(v_B) \cdot dv_B \qquad (14)$$

Hence, there is a probabilistic relationship between the time period in that the failure occurs and wear limit (V^*_B), where the end of tool life is judged by the limit of tool wear, as observed in Figure 7.

Figure 7 shows that there is a probability distribution for the observed tool wear at time t, and there is also a probability distribution for the time that the tool reached its wear limit (V^*_B). By observing the tool wear evolution, it is possible to verify that for each moment of time, there is a likelihood of tool wear being less than V^*_B. This probability is reduced when the tool usage time is increased. Beside, the time where $t = \hat{T}_1$ is defined as the mean time which the tool wear reaches limit value V^*_B. The estimated for $\hat{T}_1$ is showed in Eq. (15).

$$P(T < t) = P(v_B \geq V^*_B)$$
$$\int_0^t f(t) \cdot dt = 1 - \int_0^{v^*_V} f(v_B) \cdot dv_B \tag{15}$$

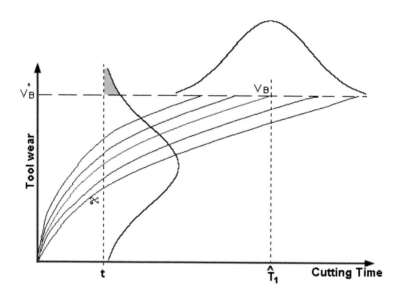

Figure 7. Relationship between time t and tool wear V_B [15].

Using the above conclusions, the tool reliability at time T=t as a wear function is presented in Eq. (16).

$$R(t) = 1 - P(T < t)$$

$$R(t) = \int_t^\infty \frac{1}{\sqrt{2\pi} \cdot (\sigma/b_6) \cdot t} \cdot e^{-\left[\frac{\ln(t)-\ln(T_1)}{\sqrt{2(\sigma/b_6)}}\right]^2} dt \tag{16}$$

Substituting $t = \hat{T}_1$ into Eq. (10):

$$\hat{T}_1 = \left[\frac{V_B^*}{C_0 \cdot s^{b_1} \cdot v^{b_2} \cdot d^{b_3} \cdot \gamma^{b_4} \cdot r^{b_5}} \right]^{1/b_6} \tag{17}$$

Thus, the tool-life distribution, which is determined from the tool-wear distribution, also obeys the Lognormal distribution as shown in Eq. (18).

$$f(t) = \frac{1}{\sqrt{2\pi} \cdot \hat{\sigma} \cdot t} e^{\left[-\frac{1}{2\hat{\sigma}^2} (\ln \hat{T}_1 - \ln(t))^2 \right]}$$

$$F(t) = \Phi \left(\frac{\ln \hat{T}_1 - \ln t}{\hat{\sigma}} \right) \tag{18}$$

4.4. Applications of Tool Reliability Analysis

Turning Process

First, the experimental data reported by Hitomi et al. [16] employing carbon steel as the cutting tool material are presented. The work piece length was 500 mm, with a diameter of 100 mm. The work piece was clamped in a high-speed lathe and a throwaway-type carbide insert tip was used to perform dry cutting. The cutting conditions are presented in Table 3. Hitomi et al. [16] compared the characteristics of reliability derived theoretically from the tool life and those obtained experimentally form the tool wear distribution. The tool reliability is kept close to maximum reliability up to 25 minutes to wear limit of 0.3 mm and 32 to wear limit of 0.4 mm, after this time, the tool reliability decreases rapidly.

Table 3. Cutting conditions Hitomi's experiment [16]

Cutting speed (m/min)	Feedrate (mm/rev)	Depth of cut (mm)	Limit value of flank wear mm
175	0.2	1.5	0.3
175	0.2	1.5	0.4

Table 4. Cutting conditions Wang's experiment [17]

	Cutting speed (m/min)			
	88.1979	125.35	208.92	313.37
Feedrate (mm/rev)	0.2	0.2	0.2	0.2
Depth of cut (mm)	1.5	1.5	1.5	1.5
Time mean tool life, T_1, (min)	245.11	84.07	17.23	4.90
Variance σ^2	3,3954	3,3395	0,3490	0,3104

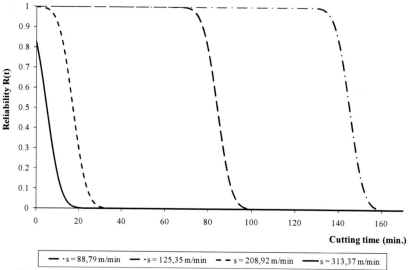

Adapted from Wang et al. [16].

Figure 8. Turning cutting tool reliability.

Wang et al. [17] developed experiments using high carbon steel as the experimental material, and a number of tool wear experiments were carried out on a heavy duty lathe employing a sintered carbide insert to perform dry cutting. The work piece length was 350 mm with a diameter of 66.5 mm. The mean time for life tool considering the limit value of flank wear as 0,3 mm was determined. The tools were tested 4 cutting speed. The experimental conditions are presented in Table 4.

Based on the theoretical development shown in section 4.3 and using the Eq. (17), the time mean tool life for each cutting speed, as shown in Table 4 was found. Figure 8 shows the reliability behavior in terms of the cutting speed, following a Lognormal distribution with mean $\hat{T}_1$ and variance.

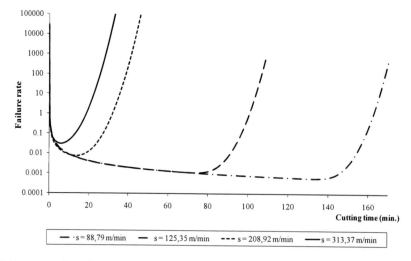

Figure 9. Failure rate of turning cutting tool.

Table 5. Cutting conditions Patino Rodriguez's experiment [15]

Diameter (mm)	5
Depth of cut (mm)	25
Feedrate (mm/rev)	0.025
Rotation (rpm)	4000
Cutting seep (mm/min)	62831

Using Wang et al.'s experimental data, the failure rate for was calculated applying Eq. (3) and presented in Figure 9. The failure rate increases for all cutting conditions. When cutting speed is low, the early failure period increases. This behavior can be explained due to the fact that when the cutting speed is slow the wear rate is lower. The tool premature failures under these conditions are due to problems with tool manufacturing process or tool material.

Drilling Process

Patino Rodriguez [15] defined experimentally the drilling tool reliability through the execution of controlled drilling test. A M2 HSS drill was used to drill holes in an ASTM 1010 steel block. During the experiments the machining feed rate and speed were controlled. The cutting conditions are presented in Table 5.

After the execution of a group of five holes, the drill flank wear was evaluated based on images captured using a microscope. The holes are drilled until the flank wear is greater than 0.120 mm, defined as the maximum allowable drill wear. That experiment was executed with 10 drills. Based on the distribution of number drilled holes until the flank wear is 0.120 mm, the cutting tool reliability is defined.

The drilling tool reliability distribution parameters are estimated through the Lognormal distribution with mean 10.08 and variance 0.0599, where the mean and variance from a domain of Normal distributions (See Figure 10).

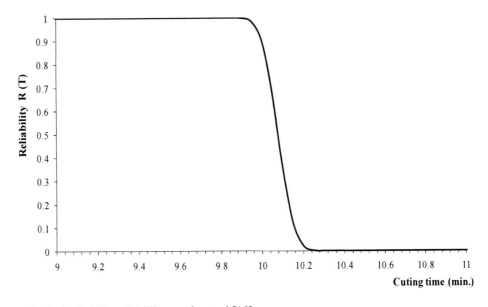

Figure 10. Tool reliability of drilling cutting tool [15].

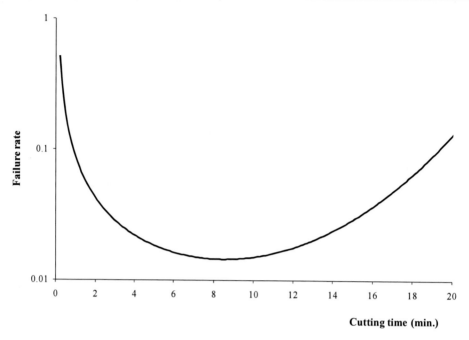

Figure 11. Failure rate of drilling cutting tool.

The drilling tool reliability decreases rapidly when the drilling time over-spends 8.40 minutes. In this case, the flank wear presents great magnitude affecting the quality of the produced holes.

The failure rate for experimental data of Patino Rodriguez [15] was calculated applying the Eq. (3) and is presented in Figure 11.

In both cases presented, it is emphasized that a Lognormal distribution is an appropriate density probability function used to model wear-related phenomena that results in failure rate increases with time, because the time to failure related to cutting tool wear usually presents a large uncertainty [7].

The tool wear has lead to the loss of dimensional and geometrical tolerances of machined parts but the relationship between the tool wear and loss of tolerance manufactured part is not yet well-defined.

4.5. Calculating Reliability of a Part Manufacturing Process

A manufacturing process is defined as a sequence of pre-established operations, aimed at producing a specific part or product. The reliability of the process may be evaluated using reliability concepts such as block diagram and probability distribution function to describe the likelihood of failures for both machines and tools. Usually, the reliability of an operation depends on three independent factors: operator, machine-tool and cutting tool. The process reliability depends on the operation sequence and their reliability. The operation reliability is a statistical relation between cutting tool, operator and machine-tool reliability.

The reliability of a part manufacturing process is mainly determined by the cutting time for each job and by the sequence of operations, defined by the series configuration. The reliability of a given job in any manufacturing process is calculated according to Eq. (20):

$$R_{operation}(t) = R_{operator}(t) \cdot R_{machine}(t) \cdot R_{tool}(t) \tag{20}$$

where, $R_{operation}(t)$ is the reliability of the manufacturing operation at a given time t, $R_{machine}(t)$ is the reliability of the machine at a given time t, $R_{tool}(t)$ is the reliability of the cutting tool at a given time t, and $R_{operator}(t)$ is the reliability of the machine operator at a given time t.

The reliability of the cutting tool may be estimated based on literature review or experimental results as we explain below. The reliability for a specific machine must be experimentally evaluated based on 'time to failure' database, as shown in 1.3 .4 the machine-tool reliability is modeled according to an exponential distribution, which the parameter is dependent on the electrical motor, spindle and bearings failure rates. The operator reliability is also defined experimentally based on the register of the number of errors that occurred during a specific period of observation.

The failure rate for a manufacturing process begins with a decreasing failure rate, named as the early failure period (also referred to as infant mortality period). Failures during infant mortality are highly undesirable and are always caused by defects and blunders: material defects, design blunders, errors in assembly or errors in tool selection. Appropriate specifications and adequate design tolerance can help, and should always be used, but even the best design intent can fail to cover all possible interactions of components in operation. In addition to the best design approaches, stress testing should be started at the earliest development phases and used to evaluate design weaknesses and uncover specific assembly and materials problems. These errors cause the production of defective parts early in the process attracting the attention of the operator and quality inspectors that there is a problem with any process or machine. This decreasing failure rate typically lasts several weeks to a few months.

Next, a useful life period is expected, where the failure rate function is constant over time. This period of constant failure rate is known as the random failure period. Useful life failures are normally considered to be random cases of "stress exceeding strength", whether the part design and processes are well designed. The failures are usually due to forces external to the product, such as mishandling, external interface failures, or accidents. However, many failures often considered normal life failures are actually infant mortality failures. The constant failure rate is used for quality control to predict the total of defective parts in a batch, and sometime in probabilistic judgments, which are based on binomial, Poisson or exponential distribution.

Concluding with a wear-out period that exhibits an increasing failure rate, which begins to increase as tool materials wear out and machines degradation failures occur at an ever increasing rate.

The analysis presented here shows that machines scheduling and tool change are essential to guarantee the reliability of a specific part manufacturing process, so tool change time is a function of the tool reliability, and the tool reliability is calculated based on tool wear. The cutting tool reliability is critical for any manufacturing process once the time to failure of the cutting tool is extremely lower than the time to failure of machines and operators, thus planning the manufacturing process based on reliability analysis allowed to optimize machining processes with respect aiming at improving product quality, reducing machine down time and lowering production cost.

4.6. Manufacturing Process Reliability Analysis and Tool Change Time Based on Reliability

A process plan generally consists of two parts. A two-layer hierarchy is considered suitable to separate the generic data from those machining-specific. The fist level focuses on part data, analysis machining feature decomposition, machining process selection, grouping of processes into jobs and decisionmaking on heat treatment. The second level considers the detailed working steps for each machining operations, including machine, cutting-tool and fixture selection, cutting-parameter assignment, machining operations sequencing and operations standard time.

The reliability of the manufacturing process is based on process planning knowledge and specifically on sequencing the operations.

Accetturi [20] proposes a model to optimize the selection of operational references, machining methods, machine, cutting-tool and fixture selection and the arrangement of machining operations sequence. Although the proposal is broader, it is only used to select the machine, cutting-tool and fixture, because the aim of this work is static process planning and the reliability is not considered as a constraints factor in the sequencing of the operations.

Accetturi [20] selects machine tools, tooling and fixtures for each machining operation. According to the proposed model, the system's adaptation to a new manufacturing cell is achieved by building a new knowledge base with rules corresponding to the current machining strategy, containing information on the updated machines, tools and fixtures. The factors that have influence on the choice of a manufacturing process are: quantity, complexity of form, nature of material, size of part, section thickness, dimensional accuracy, cost of raw material, possibility of defects and scrap rate. This step is an input for sequencing the operations.

To realize the sequencing the operations, the present authors propose to apply algorithms to obtain a sequencing of operations as a function of priorities imposed by technical, economical and geometrical constraints. Halevi [3] defines these different types of constraints. Required inputs to this phase include a description of the processes as well as machines-tools and tools needed to produce the different features with the precision and surface finish required. These inputs are analyzed and evaluated in order to select an appropriate sequence of processing operations based upon reliability concepts. This step is to define a systematic method to select a suitable process plan based on reliability concepts. The input of this step is the information about the machining features.

A task of the process planning is to select suitable machining sequences for the machined features of the parts to be machined. This process is carried out for each feature, and the following procedure is recommended:

1. To obtain of technical, economical and geometrical constrains
2. To define precedence relationships or anteriorities
3. To select the suitable machining sequence of the machined features.

The sequencing of operations can be modeled as a set of jobs in a series configuration. From the point of view of reliability analysis, the sequence of jobs can be analyzed by a series diagram block as shown Figure 12.

Hence, the sequencing of job plan processes allow the determination of the operating time of each tool and each machine to manufacture a part, and determine the job reliability as indicated in Eq. (20). Knowing the operation reliability, it is possible to establish the manufacturing process reliability as shown in Eq. (21).

$$R_{process}(t) = \prod_{i=1}^{n} R_{operation_i}(t_i) \qquad (21)$$

where $t = \sum_{i=1}^{n} t_i$, and t_i is execution time of each job.

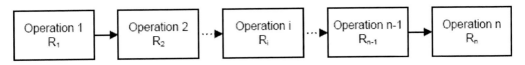

Figure 12. Block diagram system for machining operation.

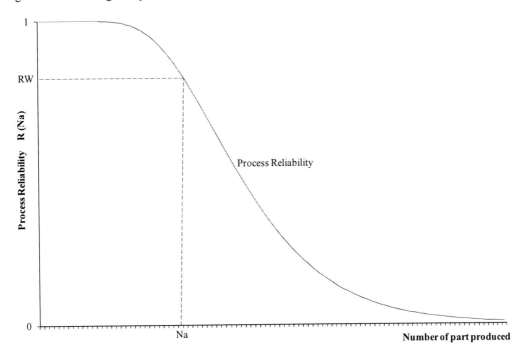

Figure 13. Reliability target in function of number of parts produced.

The operation reliability depends on machine-tool, tool, and operator reliability, nevertheless, the failure tool presents that failure rate increasing and its periodic replacement may alter the manufacturing reliability.

When an active tool wears out, it is replaced by a new one, and the consecutive time interval of tool change varies in terms of the number of parts produced in such an interval (N_a). There is a relationship between time t and N_a as shown in section 4.3. Using this relation and assuming that the reliability for each job is known, it is possible to determine the relationship between N_a and process reliability, as presented in Figure 13. If the cutting tool is not changed after N_a parts are produced, the reliability of the job is reduced, hence the number of non-conforming parts is increased. To avoid this increase it is necessary to define a target reliability required so that the tool wear is not excessive.

Assuming that the time of the job i is t_i and that the tool is changed after the production of N_a parts, cutting tool change time (Tc) for the job i is shown in Eq. (22):

$$Tc = t_i \cdot N_a \tag{22}$$

The problem of tool change time definition becomes more complex when a manufacturing process is analyzed. The manufacturing process is a series system composed of i components (jobs) [15]. In this way, assuming that the time t_i is used for each job, the reliability of the manufacturing process after N parts production is calculated as indicated in Eq. (23).

$$R_{process}(t) = \prod_{i=1}^{n} R_i(N \cdot t_i) \tag{23}$$

where $R_{process}(t)$ is the process reliability, R_i is the ith job reliability, n is the number of jobs in manufacturing process and N is the number of parts produced with the same tool [20].

If the required process reliability is Rw, the number of parts that can be produced with the same tool in each of the jobs is Na, as shown in Figure. 13.

Using the reliability of each job and assuming that each one of these operations uses different tools, the critical tool is selected and the change time for each critical tool is determined. If the manufacturing process is composed of only one job and the minimum required process reliability is Rw, the number of produced parts (Na) is calculated based on the process reliability curve. If the production required is greater than Na, the cutting tool must be changed in order to increase the manufacturing process reliability above the minimum value. This procedure is repeated until all parts required by the production planning are produced.

If the manufacturing process is composed of more than one job, the reliability block diagram for the process must be developed. That diagram is a series system composed of jobs.

This chapter considers that during the production period, reduction in both operator and machine-tool reliability are smaller than the reliability reduction caused by tool wear. The algorithm for critical tool selection is described below and is summarized in Figure 14. The use of the information of the priorities of Halevi et al., which aids on machine and tool scheduling problems, and minimizes delays in tool change, is necessary to include the information of change time for each of the tools involved in the process. In the algorithm description, the following notation is used:

T:	Total machining time of one piece $T = \sum t_i$
t_i:	Machining time of ith job
K:	Number of jobs required in manufacturing of one part
N:	Number of required parts
$R(t)$:	Process reliability (calculated)
$R_i(t)$:	i^{th} job reliability
Rw:	Minimum required process reliability (indent left)
H:	Number of tools changes.
$Tc\ (j)$:	Change time of tool j
$\lambda_i(t)$:	Failure rate of ith tool at time t

The reliability of each job is calculated based on Eq. (25) and the procedure presented in Figure 18 is applied.

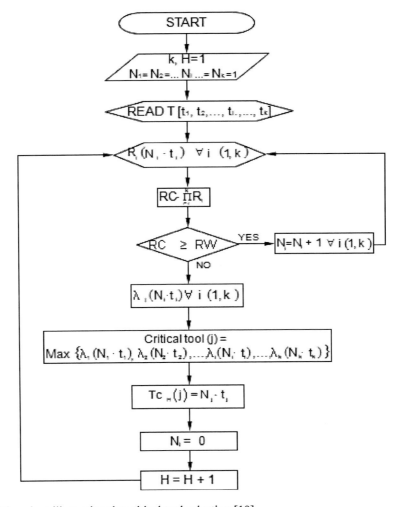

Figure 14. Flowchart illustrating the critical tool selection [19].

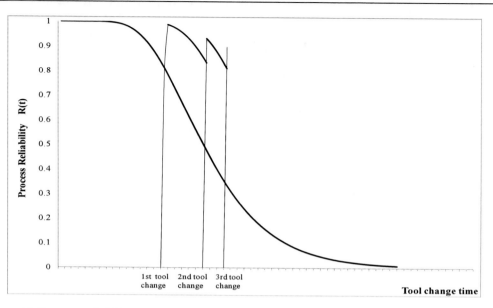

Figure 15. Process Reliability after each tool change.

The procedure basic steps are:

1. The production planner must know the production time for each job, the number of jobs used to produce the part, the total production time for one part and the number of parts to be produced;
2. The planner must calculate the reliability of each job and the process reliability after producing one part;
3. Step 2 must be executed after the production of each part and the counter related to the number of produced parts, the process reliability and the cutting time for the tools used in each job must be continuously updated after the production of one part;
4. The process reliability must be compared with the minimum required reliability at the end of the production of each part. If that reliability is greater than the minimum value, another part can be produced. On the contrary, the manufacturing job with the lowest reliability must be identified and the cutting tool used in the job must be changed. That cutting tool is named as the critical cutting tool;
5. The cutting time for the changed tool is set equal to zero and its reliability is restored to 1. The manufacturing job reliability is increased after tool change. After that change the manufacturing process reliability must be re-calculated;
6. The analysis must return to Step 2. The analysis continues until the number of produced parts achieves the value required by the production planner.

The method to establish tool change time is dynamic since it considers the process reliability based on tool wear for each job. This reliability represents the probability that a given tool reaches the target wear level at a specific time. This target wear level must be selected based on literature information and represents the maximum admitted wear for the tool without affecting the tolerances required for the part being manufactured. The change time of each tool is based on the analysis of the tool hazard rate, guaranteeing that the cutting tool is replaced before it causes a loss of manufacturing parts capability. The algorithm is flexible once it does not pre-define the tool change time. This time is changed during the manufacturing process depending on the condition of the each feature and each tool. After each tool replacement, the process reliability is changed and improved according to Figure 15.

5. APPLICATION

This section presents two applications of manufacturing process planning analysis.

The first example presents the reliability analysis of a precision drilling process, defining as failure, the production of holes with dimensional or geometrical tolerances out of the range defined by the part designers. The reliability analysis is based on the application of Failure Modes and Effects Analysis (FMEA) to analyze the drilling process, in order to define the consequences of machine tool and tool failures, and even drilling conditions out of specification, on the process. The consequences are expressed as a loss of the part tolerances.

The second example discusses the application of the manufacturing process reliability concepts and reliability-based tool change time method in a machining process planning.

Figure 16. High speed precision drilling machine.

5.1. Precision Manufacturing Process Failure Mode and Effects Analysis

A precision manufacturing process is defined by the dimensional and geometrical tolerances that can be achieved by the manufacturing process. The higher the precision, the greater are the costs associated with the process. Precision manufacturing costs are usually used for mechanical parts that have a great responsibility in providing the structural stability for a given mechanical system. The structural stability is associated with low clearance between parts, absence of surface flaws, or controlled surface roughness. The excessive clearance between structural members can cause wear of fasteners or other joining elements in structures subjected to dynamic loading. Also for dynamic loaded structures, the presence of flaws or excessive surface roughness can increase the fatigue crack growth process.

For aircraft, structural parts that are usually joined with rivets, the hole for rivet installation must be precisely manufactured to provide structural stability. The precision is associated with the dimensional and geometrical tolerances defined for the hole. For aircraft structural parts, the tolerances are very tight and must be designed to be joined with regular size rivets. To achieve the design tolerances, those holes must be manufactured using a precision drilling process.

Taking in view the dimensions and geometry of some aircraft structural parts, the drilling machine must be portable. Furthermore, in order to achieve the high cutting speeds recommended for drilling, some materials employed on those structural parts pneumatic motors power those machines. A typical high speed-drilling machine is shown in Figure 16. Those machines also present a high rotation precision and a device for their coupling to a jig used as a drilling mask.

Table 6. Failure modes and effects analysis for a precision drilling process

a. Precision Drilling Process

Function	Failure Modes	Cause of Failure	Possible Effects	Detection	Criticality
Drill a hole of a pre-defined diameter, with controlled dimensional and geometrical tolerances.	Hole dimensional and form errors.	-Drill wear; -Buckling of the drill; -Drill vibration; -Machine tool spindle run-out; -Inadequate chip removal.	-Reprocessing, in case of dimension smaller than the lower limit; -Use of non-standard rivets, for holes larger than the upper limit.	Dimensional control.	High, for diameters larger than the upper limit.
	Hole perpendicularity errors.	-Incorrect machine tool set up; -Spindle misalignment; -Incorrect drill fixture; -incorrect jig set up.	-assembling problems; -reprocess.	Measurements with a specific metrology procedure.	High
	Hole surface defects.	-Drill wear; -Drill vibration; -Inadequate chip removal.	-Excessive roughness; -Presence of flaws.	Roughness measurement.	Very high
	Hole center position error.	-Incorrect jig manufacturing.	-Assembling problems.	Measurements with a specific metrology procedure	Very high

b. Drill

Function	Failure Modes	Cause of Failure	Possible Effects	Detection	Criticality
Drill a hole of a pre-defined diameter, with controlled dimensional and geometrical tolerances	Wear	-Incorrect cutting speed and feed rate; -Defects in raw material; -Lubrication problems; -Incorrect drill material selection.	-Hole dimensional, location and form errors; -Hole surface defects	-Cutting force or torque monitoring -Flank wear measurement	Very high
	Fracture	-Wear mechanism progression; -Thermal or mechanical fatigue; -Refrigeration and lubrication problems; -Incorrect tool material selection.	-Hole dimensional, location and form errors; -Hole surface defects	Visual inspection	Very high

The aircraft part for which the drilling process is analyzed in this section is located on the plane's main body. The holes are used to install rivets to join two structural sections. The rivets are standard parts and any dimensional or geometrical defect in the holes must be corrected and may induce the use of non-standards rivets to join the structural parts. The aircraft manufacturer must inform the customer about any change in the aircraft components. In future scheduled maintenance actions the maintenance teams must have the non-standard part to be installed in the aircraft. Those changes can increase the aircraft maintenance costs and the complexity of the maintenance planning.

In order to guarantee the correct location of the holes in the part, the manufacturing process employs a special jig, clamped on the part surface with bolts that defines the holes drilling position.

The pneumatic drilling machine uses the holes in the jig as a datum. The chuck's front part of the drilling machine expands, clamping the drilling machine on the jig datum hole's internal surface. This operation guarantees the correct location of the holes and also helps to keep the perpendicularity of the hole regarding the part surface.

Once the drilling machine is installed on the fixture, the drilling process starts. The drilling plan defines the cutting speed and the feed rate that must be adjusted by the operator. The tools are a special step drill that performs the step drilling and the reamer for finishing process, necessary to fulfill the dimensioning and tolerance requirements for each hole. Once the set up is completed, the machine operation is fully automatic.

Considering the drilling process as a system that must be capable of performing a function for a given period of time, the failure mode and effects analysis can be employed for enumerating the possible modes by which the drilling process components, such as drilling machine, drilling tool and even operator, may fail and for tracing through the characteristics and consequences of each mode of failure on the process as a whole.

The failure modes and effects analysis for the precision drilling operation is presented in Table 6.

The criticality associated with each failure mode is related to the consequences of the failure mode on the manufacturing process capability, and consequently, on the hole's dimensional and geometrical characteristics. This attribute is used to separate failure modes that are very critical from those that merely cause inconvenience or moderate economic loss.

As for the precision drilling operation, the surface roughness is a very important control parameter, once the presence of high roughness can increase the probability of fatigue failure of the part. Due to dynamic loading, a crack growth can be foreseen during the aircraft life, which can induce unscheduled maintenance actions, increasing the aircraft operational costs. The maintenance actions involve the crack repair increasing the hole size. Consequently, a non-standard rivet must be used which also increases future maintenance costs.

Any error in the hole center position is very critical for the assembly process. The correct position of the hole center is defined by the drilling jig that can be considered a precision manufactured part. The jig is fully inspected and any imperfection must be corrected before the part is sent to manufacturing plant. Although the jig is considered perfect, any misalignment in the jig mounting can cause problems in the hole center position and even in the perpendicularity of the hole centerline regarding the part surface. So a detailed inspection procedure and special fixture devices must be used when mounting the jig on the part to be drilled.

The hole dimension can be considered critical as for rivet mounting in the assembling line. If the drilled hole is smaller than the lower dimensional limit, a correction action must be taken involving the reprocessing of the hole. This action increases the manufacturing cost but fixes the manufacturing error. Therefore, if the hole dimension is higher than the upper dimensional limit, a non-standard rivet must be used in the assembling line, increasing the future aircraft maintenance costs, due to the use of non-standard parts.

Based on the FMEA analysis results, the process planner must carefully design the manufacturing sequence in order to minimize the probability of occurrence of those defects. The planning tasks involve the drill geometry and material selection, the machine tool selection and the drilling conditions, such as cutting speed and feed rate.

Even using a perfectly designed manufacturing process; the drilled hole presents some dimensional and geometrical variability, which is associated with the process reliability. Manufacturing process reliability can be represented by the bathtub curve presented in Figure 2. The process reliability can be defined as the probability that the dimensional and geometrical characteristics of the manufactured part respect the design tolerance. The manufactured part geometry and dimensions are related to the tool wear [8]. The process life can be modeled as the time to wear a new drill until it fails by excessive wear. The failure rate is related to the probability of drilling holes with dimensions or geometry out of the design's acceptable limits.

In the initial stages, modeled as the first drilled holes, a decreasing failure rate is expected. These infant deaths are related to improper machine tool set-up, use of incorrect drill material or even improper jig mounting. Those failures are easily detected through inspection methods, as those presented on the FMEA table. Once the defects are detected, a corrective action is taken, in order to eliminate the cause of failure. The middle section of the bathtub curve contains the smallest and most nearly constant failure rate. Failures during this period of time are associated with unavoidable overloading in the tool due to deviations in the material mechanical properties, vibrations in the machine tools, machine resolution, repeatability and accuracy, or even random human errors. On the right of the bathtub curve is a region of increasing failure rates. During this period of time, aging failures take place, related to the cumulative wear of the drilling tool, which is the most important cumulative tool failure mechanism. The consequence of the increasing failure rate in a great increase in the production of holes out-of-specification, increasing the need for re-processing, once the tool wear affects the diameter of hole.

The tool wear is a cumulative process that starts with the first use of the tool. In the beginning of the tool life, the wear is small, and the failure rate associated with that failure mechanism is very small in comparison with the early failure rate or even to the random failure rate. The wear grows with the use of the drilling tool and also the probability of drilling holes out-of-specification, increasing the failure rate.

Although the tool wear can be taken as a tool failure indicator, the mechanical manufacturers usually define a tool life based on empirical observations and not on reliability concepts. In order to avoid errors during the manufacturing process, quality control techniques are used to evaluate the manufactured parts. Although the quality control techniques based on control charts and continuous process capability analysis are used to check if the parts are manufactured properly, they are not directly related to the process reliability as discussed in the present chapter.

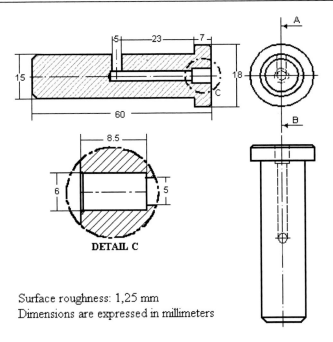

Figure 17. Mechanical drawing of the rotor bearing with lubrication hole.

Table 7. Characteristic for each feature

Group	Machining Operation	Stage	Specifications ɸ [mm]	L [mm]	Cutting conditions [11], [16] Feed Rate [mm/rev]	Depth [mm]	Speed [m/min.]	Time [min.]
R_1	Drilling		5	7.5	0.025		62.83	0.0750
	Turning	Rough	18.3	60	0.15	0.5	125.35	0.1835
R_2	Turning	Rough	15.3	53	0.15	0.5	125.35	0.4342
	Turning	½ Finish	18.1	60	0.1	0.1	208.92	0.1633
R_3	Turning	½ Finish	15.1	53	0.1	0.1	208.92	0.1203
	Turning	Finish	18	60	0.08	0.05	313.37	0.1353
R_4	Turning	Finish	15	53	0.08	0.05	313.37	0.0996
	Drilling		6	8.5	0.025	40	62.83	0.1020
R_5	Drilling		5	23	0.025		62.83	0.2300

5.2. Reliability of Manufacturing Process

A case study is used to illustrate the application of the methodology to define tool change time. The case study uses the machining process of a shaft with a lubrication hole used to distribute lubricating fluid to the bearing. The shaft is machining from an SAE/AISI 1010 steel cylinder, with length of 60 mm, and diameter of 18.5 mm (See Figure 17). The specific dimensions, forms, tolerances, and roughness required are obtained using turning and drilling operations.

Table 7 presents dimensions, specifications of machining and the order of precedence and machining times for each operation. Precedence operation is calculated using algorithms and the models presented by Patino Rodriguez [15]. Data presented in this table allows the manufacturing planner to obtain the necessary information about the part.

Reliability calculations are somewhat limited by the available data in the literature and the following hypotheses are used to calculate process reliability.

Consecutive operations are grouped in a same block when the following requirements are fulfilled:

- Machining conditions are identical during operation,
- The machining-grouped operation uses the same tool and,
- The operation of the block is machining in consecutive order (one after the other r).

The tools used for the *finishing*, *½ finishing* and *rough* operations are different, although the machines used in the process (lathes and drilling machine) have an equal hazard rate. In this case, and according to the results presented by Wang [12], the reliability of the machines is modeled using exponential distribution, where the 'mean time to failure' is equal to 43603.8 minutes [12]. The mean time between failure of operators involved in the process is much longer than the manufacturing time for the parts, assuming that during the time of part's manufacture the reliability of the operator is equal to one ($R_{operator}=1.0$).

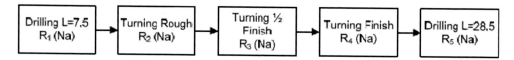

Figure 18. Block Diagram System for machining shaft with lubrication hole.

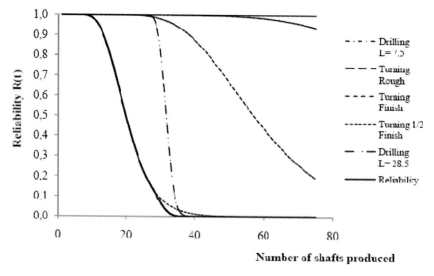

Figure 19. Reliability for manufacturing process without tool change in function of number of shafts produced.

Table 8. Lognormal distribution parameters for cutting tool reliability analysis

Operation	Tool*	T_l	σ^2
Drilling	Drill	10,08	0.0599
Turning Rough	Insert	84.07	0.3395
Turning ½ Finish	Insert	17.23	0.3490
Turning Finish	Insert	4,90	0.3104

Drill: M2 High Speed Steel [16]. Insert: Inserted carbide tip, TNMG160404L2G [11].

Figure 18 shows the block diagram for this case study, according to data presented in Table 7. These blocks represent the part manufacturing operation with a machining time equal to the addition of the machining times for the each operation.

The reliability of the turning and drilling operations is modeled as shown in section 4.4.

Table 8 shown the Lognormal distribution parameter used to model cutting tool reliability where μ and σ^2 are the mean and variance from a domain of normal distributions.

Reliability for the manufacturing process for one part produced is calculated substituting the Eq. 20 in Eq. 21 considering the reliability parameters for machines, tools and operators:

$$R_{process} = \left[\left(1 - \Phi\left(\frac{10.08 - t}{\sqrt{0.0599}}\right)\right) \cdot \left(e^{4360.8^{-1}t}\right) \cdot (1)\right]$$
$$* \left[\left(1 - \Phi\left(\frac{84.07 - t}{\sqrt{0.3395}}\right)\right) \cdot \left(e^{43603.8^{-1}t}\right) \cdot (1)\right] *$$
$$* \left[\left(1 - \Phi\left(\frac{17.23 - t}{\sqrt{0.3490}}\right)\right) \cdot \left(e^{43603.8^{-1}t}\right) \cdot (1)\right] * \left[\left(1 - \Phi\left(\frac{4.90 - t}{\sqrt{0.3104}}\right)\right) \cdot \left(e^{43603.8^{-1}t}\right) \cdot (1)\right]$$

The time to produce one part, without considering setup time, is 1.562 minutes, and reliability for one shaft manufacturing process is 0,99996. Figure 19 shows the reliability behavior of each manufacturing operation in terms of the number of parts. Turning the ½ finish tool causes the decrease in process reliability for two reasons: this tool has a high hazard rate and it is in use for the longest time. The combination of these two factors implies that this tool has the shortest change time.

This information allows for the calculating of the process reliability and to apply the algorithm presented in Figure 14. Through the increase in the number of manufactured parts, the process reliability is continuously recalculated, once that reliability becomes lower than a given target value, in the present case 80%, a tool must be changed aimed at restoring the process reliability. The tool to be changed is the one with the lowest reliability. The tool change will restore that cutting tool reliability to one.

The process reliability is recalculated and the procedure presented in the preview paragraph is repeated until the reliability is lower than 80% and another tool change is necessary.

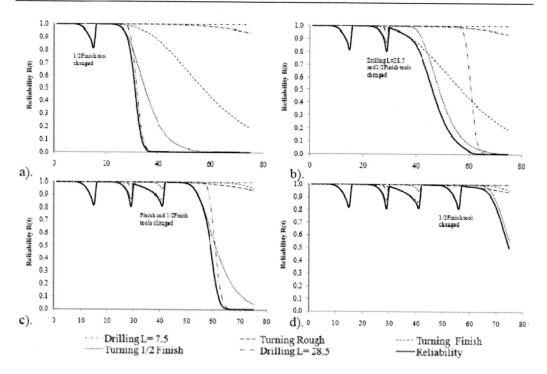

Figure 20. Reliability for manufacturing process with tool change as a function of number of shafts produced. a). 1st tool change. b). 2nd tool change. c) 3^{rt} tool change. 4th tool change.

The value of reliability for the process is reduced until the manufacture of part 16, where the process reliability reached the value of 75.70%, is lower than the required minimum reliability. This value indicates that after of the manufacture of the 15th part, the tool that presents the greatest hazard rate or minor reliability in that instant must be determined. The tool that presents the greatest failure rate, after the manufacturing of 15 parts, is turning the ½ finish tool. Then it is the critical tool and it must be changed. After the tool change, the process reliability is calculated again, and it is observed that the reliability of the manufacturing process for part 16 is 99.94%. Here we can observe the positive effect of this tool change on the process. As the value of the reliability is superior to the required minimum reliability value, this means that the other tools also do not need to be changed, and the process can continue to run until the reliability reaches a value less than 80% (See Figure 20).

The process reliability when manufacturing the shaft, with the changes required to keep the process reliability within the desired reliability limits (0.80 – 1.00) is shown in Figure 20. This figure shows that the number of tools changed, according to the number of shafts produced. It also shows that the tool wear affects the reliability of the manufacturing process.

6. CONCLUSION

The manufacturing process planning is a very important activity in any industrial process. For mechanical industries, the main goal of manufacturing process planning is to define the sequence of operations, machines tools and tools that will be used to manufacture the parts of a given mechanical equipment.

The capacity of the manufacturing process defined experimentally, used to define the chance of manufacturing non-conforming parts. The manufacturing of non-conforming parts is usually caused by failures of the machine-tools or even by non-expected deterioration of the tools. The prediction of the long term performance of machine tools and tools can be based on the reliability concept.

The interaction between machining process parameters, machine reliability and tool reliability has been presented in order to development a method to integrate those concepts to the manufacturing process planning activity. The conditions of the process as defined by the production plan and the machining system influence the process reliability. To the process planner, it is evident that the modeling of the machine and tool reliability can avoid the production of non-conforming parts.

The experiments have shown that Lognormal distribution is the most representative distribution for the tool wear behavior. This model is particularly useful for failure processes which result from many small cumulative factors.

The reliability function for the manufacturing process was derived using tool wear distributions, and machine failure distributions. The calculation of the tool reliability was based on the maximum limit of tool wear and minimum target level for process reliability.

The application process manufacturing reliability avoids that the machining operation reliability decreases that affects the manufacturing process capability, resulting in an increase in the production of non-conforming pieces. The process reliability estimative can be used to define the sequence of tool change and the frequency of that change.

REFERENCES

[1] Savsar, M. *Reliab. Eng. Syst. Safe*. 2000, 67, 147-152.
[2] Adamyan, A.; He, D. *Reliab. Eng. Syst Safe*. 2002, 76, 227-236.
[3] Halevi, G.; Weill, R. D. Principles of Process Planning; Chapman and Hall: London, 1995; pp 15-33.
[4] Groover, M.; Fundamentals of Modern Manufacturing. Prentice-Hall: London, 1996; pp 449-572.
[5] Rozenfeld, H.; in Proceeding Symposium on CAE/CAD/CAM. Sociedade Brasileira de Comando Numérico, Sao Paulo, 1989.
[6] Rezende F.; Planning Process Computer-Aided Manufacturing Through an Expert System-Based Technology Features: A Model for Developing Facing Reality Industrial; Master of Engineering Thesis presented at Mechanical Engineering Graduate Program: UFSC (in portuguese).
[7] Lewis E.; Introduction to reliability engineering. Wiley: New York; 1987; pp 138-163.
[8] Furnee, R. J.; Wu, C.L.; Ulsoy, A.G. *J. Manuf Sci. E-T ASME*. 1996, 118, 367-375.
[9] Modarres, M.; Reliability and Risk Analysis, What Every Engineer Should Know About; Marcel Dekker: New York, NY, 1993; pp 67-137.
[10] Meeker, W. Q.; Escobar, L. A. Statistical Methods for Reliability Data. Wiley: New York; 1998; pp 1-25.
[11] Kecklund, L. J.; Svenson, O. *Reliab. Eng. Syst. Safe*. 1997, 56, 5-15.
[12] Wang, Y.; Jia, Y.; Yu, J.; Zheng Y.; Yi S. *Reliab. Eng. Syst. Safe*. 1999, 65, 307-314,.

[13] Mejabi O.; Black J. *Reliab. Eng. Syst. Safe.* 1995, 48, 11-18.
[14] Freiheit T.; Hu J. *J. Manuf. Sci. E-T ASME.* 2002, 124, 296-304.
[15] Patino Rodriguez C. Reliability applied for manufacturing process of mechanical systems. Master of Engineering Thesis presented at Mechanical Engineering Graduate Program: Polytechnic School: EP-USP, 2004 (in Portuguese).
[16] Hitomi K.; Nakamura N.; Inoue S. *J. Eng. Ind-T ASME.* 1979, 101 (2) 185-190
[17] Wang, K. S.; Lin, W. S.; Hsu, F. S. *Int. J. Adv. Manuf. Tech.* 2001, 17, 707-709.
[18] El Wardanit, T.; Elbestawi, M. *Int. J. Adv. Manuf. Tech.* 1997, 13 (1), 1-16.
[19] Patino Rodriguez C; Souza, G. F. M. *Reliab. Eng. Syst. Safe.* 2010, 95, 866-873.
[20] Accetturi, G. Decision Support System for selection of operating conditions for the machining process. Master of Engineering Thesis presented at Mechanical Engineering Graduate Program: Polytechnic School: EP-USP, 1997 (in Portuguese).

Chapter 2

DYNAMIC STUDY OF A CENTERLESS GRINDING MACHINE THROUGH ADVANCED SIMULATION TOOLS

Iker Garitaonandia, Joseba Albizuri, M. Helena Fernandes, Jesús Mª Hernández and Itxaso Olabarrieta

Department of Mechanical Engineering,
University of the Basque Country, Spain

ABSTRACT

Centerless grinding operations present some characteristic features, which make the process especially prone to suffering dynamic instabilities, leading to chatter vibrations. This kind of self-excited vibrations are very pernicious, as they not only limit the quality of the machined parts, but also the lifetime of the machine and the tool.

The design of control systems adapted to reduce vibrations requires the availability of accurate models capable of evaluating the effectiveness of different control alternatives prior to their practical implementation, so an optimization process can be tackled via simulations.

In this chapter a design procedure is presented to obtain reliable low-order models of the dynamic response of a centerless grinding machine. For this purpose, an approach based on the combined use of both numerical and experimental techniques was selected. Initially, the finite element (FE) model of the machine was updated taking as reference data obtained from an experimental modal analysis (EMA). This updated model showed to accurately characterize the vibration modes excited in the machine structure under chatter conditions. Nevertheless, the major drawback is that it has a large number of degrees of freedom, implying computationally expensive calculations. Thus, in a second step, the updated FE model was reduced to obtain a low order state space model covering the dynamic characteristics of the machine in the frequency range of interest.

The reduced model was integrated in the chatter loop of the centerless grinding process, and several simulations were performed both in the frequency and time domains. Different machine responses were estimated in these simulations and the results were compared to those obtained in machining tests, achieving good agreement between the theoretical and experimental results and, hence, validating the reduced model.

1. INTRODUCTION

Centerless grinding is a chip removal machining process where the workpiece is not clamped but resting against the grinding wheel, the regulating wheel and the workblade (figure 1).

The regulating wheel, whose angular velocity is ω_r, is made from high contact friction material and is responsible for controlling workpiece angular velocity, ω_p, so their surface speeds are approximately identical. The grinding wheel is made from abrasive grains, rotates at a much greater angular velocity ω_g than the regulating wheel, and is responsible for eliminating material from workpiece. Workblade acts as workpiece support between both wheels and is height adjustable to raise or lower workpiece center in relation to the center line of both wheels.

The special configuration of this process eliminates the need to make anchoring holes in the workpieces to be machined, thus avoiding this operation and possible precision errors in centring. Thus workpieces can be obtained with highly accurate dimensional tolerances together with high production rates, since workpiece loading and unloading times are reduced to a minimum.

However, problems associated with roundness errors of the parts are very common in these machines due to the workpiece floating center. Consequently, the workpiece surface errors, after contacting the workblade and the regulating wheel, produce a displacement of its center, which under certain conditions may lead to an unstable error regeneration process.

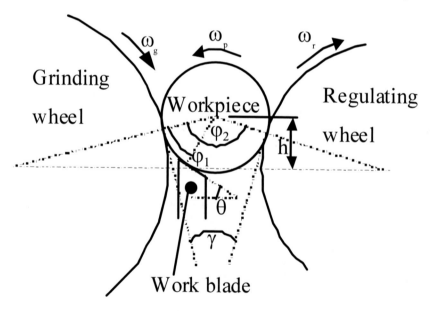

Figure 1. Geometric configuration of centerless grinding [1].

The instabilities in centerless grinding operations can be geometric or dynamic in nature [2, 3, 4, 5]. Geometric instabilities are specific to the centerless grinding process, and generate surface undulations on the workpiece due solely to the geometric configuration of the machine, regardless of its dynamic characteristics. However, dynamic instabilities generate undulations due to the interaction between the cutting process and the dynamic characteristics of the machine, giving rise to self-excited vibrations or chatter, characterised by the appearance of very violent machine vibrations.

Study of this latter problem requires knowledge of the dynamic properties of the machine. These properties have been traditionally studied adopting simplified models where the machine is characterised by one or two representative mode shapes [**Error! Reference source not found., Error! Reference source not found., Error! Reference source not found., Error! Reference source not found., Error! Reference source not found., Error! Reference source not found.**]. Nevertheless, to obtain an appropriate knowledge of the machine dynamic response, it is highly useful to have more elaborated numerical models enabling correct simulation of the structural properties of the machine being studied.

Bearing this in mind, this chapter details the characterisation process of a centerless grinding machine by means of finite element (FE) model updating techniques using data obtained from the experimental modal analysis (EMA). Subsequently, taking the updated FE model, a reduced order model is obtained capable of correctly predicting the machine chatter behaviour.

2. System Modelling

This section describes the centerless grinding modelling process. Figure 2 shows the main components of these machines.

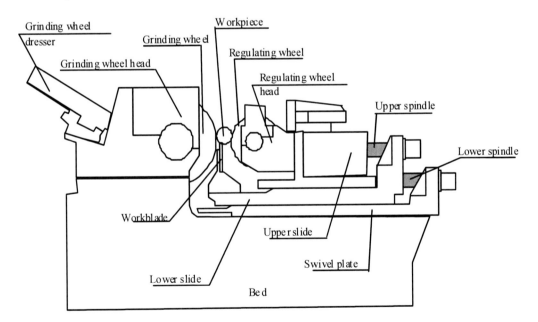

Figure 2. General diagram of centerless grinding [8].

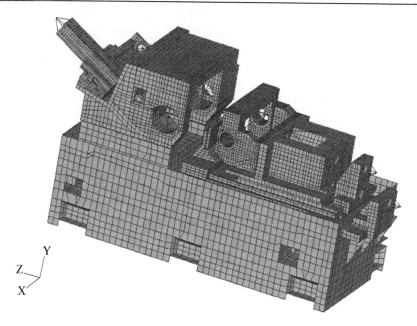

Figure 3. Centerless grinding FE model [8].

2.1. FE Model

The dynamic characteristics of the machine were studied theoretically using the FE model shown in figure 3.

This model has 53200 nodes and 37807 elements. Z axis is defined as the longitudinal axis, X as the transverse axis and Y as the vertical axis.

2.1.1. Numerical Modal Parameters

In the presence of any load applied to the centerless grinding machine, the FE model shown in section 2.1 describes the dynamic equilibrium between the external, the elastic and the inertial forces. As figure 3 shows a discrete system, the equation of motion can be expressed using the following matricial/vectorial expression:

$$[M] \cdot \{\ddot{z}(t)\} + [K] \cdot \{z(t)\} = \{F(t)\} \tag{1}$$

This equation represents a conservative (undamped) system simultaneously expressed by N ordinary second order linear differential equations with constant coefficients, N being the number of degrees of freedom (dof's) of the FE model. In it $\{\ddot{z}(t)\}$ and $\{z(t)\}$ are the vectors of generalized accelerations and displacements, respectively. $\{F(t)\}$ is the vector of applied forces. $[M]$ and $[K]$ are respectively the mass and stiffness matrices.

As a starting point for the dynamic study of a structure it is interesting to consider the free response of system (1), which is:

$$[M] \cdot \{\ddot{z}(t)\} + [K] \cdot \{z(t)\} = \{0\} \tag{2}$$

This equation may be used to obtain the mode shapes and the natural frequencies of the structure, whose knowledge is very interesting when interpreting its dynamic response.

The solution of equation (2) is expressed as:

$$\{z(t)\} = \{\psi\}_i \cdot e^{j\omega_i t} \tag{3}$$

where ω_i is the natural frequency number i of the structure and $\{\psi\}_i$ is the associated mode shape.

Substituting (3) in equation (2) leads to the following problem of eigenvalues and eigenvectors:

$$([K] - \lambda_i \cdot [M]) \cdot \{\psi\}_i = \{0\}; \quad \lambda_i = \omega_i^2 \tag{4}$$

Matrices $[M]$ and $[K]$ are symmetric; $[M]$ is positive definite (all its eigenvalues are positive) and $[K]$ is positive semi definite (all its eigenvalues are non-negative). Considering these characteristics, each eigenvalue λ_i of equation (4) is real and non-negative. Equation (4) defines only the shape, but not the amplitude of the modes, which may be arbitrarily scaled.

There are as many natural frequencies and mode shapes as there are dof's, i.e., N. The complete solution of the undamped system (2) is usually expressed by two N × N matrices:

$$[\Omega^2] = \begin{bmatrix} \omega_1^2 & 0 & \cdots & 0 \\ 0 & \omega_2^2 & \cdots & 0 \\ \vdots & \vdots & \ddots & \vdots \\ 0 & 0 & \cdots & \omega_N^2 \end{bmatrix} \tag{5}$$

and:

$$[\Psi] = [\{\psi_1\}, \{\psi_2\}, \ldots, \{\psi_N\}] \tag{6}$$

These two matrices contain a total description of the structural dynamic characteristics.

The natural frequencies defined in expression (5) indicate the frequencies to which the structure tends to vibrate. Mode shapes defined in expression (6) define a coordinate transformation which simultaneously orthogonalises the mass and stiffness matrices. This means that:

$$\{\psi\}_j^T \cdot [K] \cdot \{\psi\}_i = 0 \tag{7}$$

$$\{\psi\}_j^T \cdot [M] \cdot \{\psi\}_i = 0 \tag{8}$$

$\{\psi\}_j$ and $\{\psi\}_i$ being two different mode shapes. The orthogonality conditions (7) and (8) may be expressed in matrix form as:

$$[\Psi]^T \cdot [K] \cdot [\Psi] = [k_r] \tag{9}$$

$$[\Psi]^T \cdot [M] \cdot [\Psi] = [m_r] \tag{10}$$

where $[k_r]$ is the diagonal matrix of modal stiffnesses and $[m_r]$ is the diagonal matrix of modal masses. The relationship between these two matrices is:

$$[k_r] = [\Omega^2] \cdot [m_r] \tag{11}$$

Since the matrix of mode shapes [Ψ] can be scaled arbitrarily, [k_r] and [m_r] are not unique. Among the existing normalisation procedures, the mass-normalisation is very commonly used. In this case, the matrix of mode shapes is written as [Φ], and satisfies the following properties:

$$[\Phi]^T \cdot [K] \cdot [\Phi] = [\Omega^2] \tag{12}$$

$$[\Phi]^T \cdot [M] \cdot [\Phi] = [I] \tag{13}$$

$[I]$ being the identity matrix.

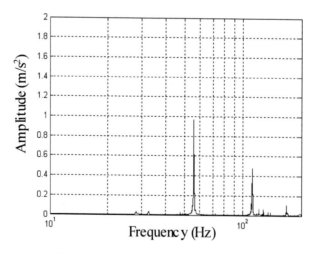

Figure 4. Frequency content of the acceleration signal measured in the regulating wheel head under chatter conditions.

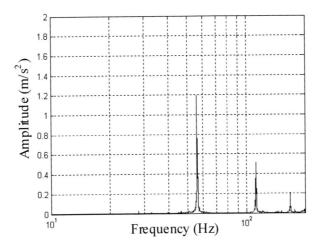

Figure 5. Frequency content of the acceleration signal measured in the grinding wheel head under chatter conditions.

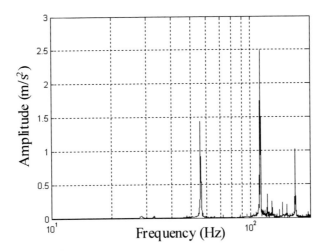

Figure 6. Frequency content of the acceleration signal measured in the workblade under chatter conditions.

2.1.2. Definition of the Frequency Range of Interest

Various machining tests were performed under chatter conditions to detect the frequencies associated to the self-excited vibrations. In these tests, the structural response was measured using a triaxial accelerometer located at various points on the machine. The analysis of the acceleration signals showed that the most important vibrations occur in the feed direction of the upper slide (longitudinal direction). Subsequently, the frequency content of the measured signals was analyzed. The results for three points located on the regulating wheel head, the grinding wheel head and the workblade are shown in figures 4, 5 and 6, respectively.

These figures show that the vibrations in the regulating and the grinding wheels are dominated by a mode shape excited at around 55 Hz, while in the workblade the vibrations are dominated by a mode excited close to 130 Hz. Among these two modes, the former is the most important, as vibrations occurring in the grinding and the regulating wheels have a great influence in the rounding errors of the workpieces.

Taking into account these chatter frequencies and with the idea of covering all the modes which might contribute to the response under chatter conditions, the frequency range of interest was established at 0 - 160 Hz.

2.1.3. Extraction of Mode Shapes and Natural Frequencies

The Block-Lanczos algorithm was applied to the FE model and a total of 15 mode shapes and natural frequencies were obtained in the 0 – 160 Hz frequency range. These frequencies are shown in table 1.

The aim of obtaining modal parameters is to determine which modes contribute most significantly to the response under chatter conditions. For this purpose, the numerical results shown in table 1 require verification if to be used later, since they were obtained from a FE model where a series of idealisations were made, which might lead to model errors.

The FE model was updated using experimental data. The interest here lies in that the data obtained experimentally provide a more real and reliable representation of the dynamic properties of the structure [**Error! Reference source not found.**]. Its application requires an experimental modal analysis (EMA) of the machine to be done.

2.2. Experimental Analysis

The EMA consists of determining the structure dynamic behaviour from the frequency response functions (FRFs) obtained experimentally. The advantage of this procedure over the numerical is that the real machine is tested, with all the possible particularities.

Table 1. Numerical natural frequencies of the original FE model

FE mode	Frequency (Hz)
1	30.16
2	31.84
3	45.9
4	54.85
5	67.38
6	73.5
7	82.27
8	93.03
9	96.84
10	102.02
11	113.28
12	122.13
13	126.82
14	149.88
15	159.44

Table 2. Test setup to obtain the experimental FRFs

Support conditions:	*In Situ*
Excitation points:	1
Excitation method:	Impact hammer with force sensor
Response points:	69
Response form:	Triaxial accelerometers
Frequency bandwidth:	0 - 200 Hz
Resolution:	401 lines

Table 3. Experimental modes

EMA mode	Freq. (Hz)	Damp. (%)	Description
1	33.41	5.8	Suspension movement in Z direction
2	48.43	3.53	Bed torsion in Z direction
3	58.91	3.5	Out of phase movement between wheel heads in Z direction
4	76.97	3.9	Bed torsion in X direction
5	90.52	2.6	Bed torsion in Z direction
6	108.44	2.6	Lower slide bending in X direction
7	122.02	1.8	Bed bending in X direction
8	128.48	1.6	Lower slide torsion Y direction
9	129.86	1.1	Lower slide bending in X direction
10	144.02	2.0	Grinding wheel dresser bending in X direction

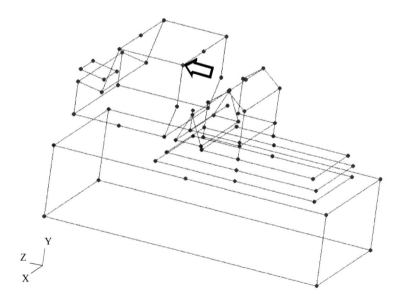

Figure 7. Machine geometry to obtain the experimental FRFs [8].

Data acquisition on the grinding machine was performed using the conditions shown in table 2.

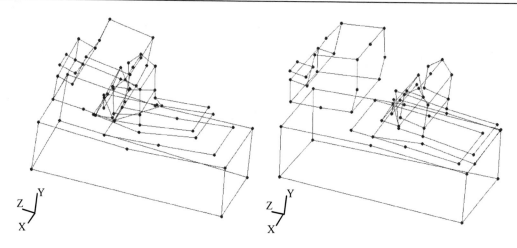

Figure 8. Main chatter mode.

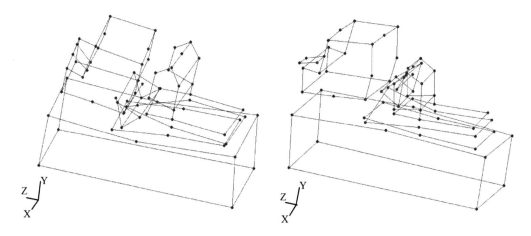

Figure 9. Secondary chatter mode.

The accelerance FRFs corresponding to 207 dof's were obtained under these conditions. Figure 7 shows the geometry used in the analysis, where the arrow indicates the point and direction of excitation. This excitation direction was chosen to excite the modes with high modal displacement content in the upper slide feed direction (Z direction).

From the measured FRFs, the system modal parameters were extracted. To this effect, the MDOF estimation technique was used in time domain, obtaining the natural frequencies and modal dampings via the LSCE method (*Least Squares Complex Exponential*) and the mode shapes via LSFD (*Least Squares Frequency Domain*). 10 natural frequencies, mode shapes and damping factors were obtained in the frequency range of interest. Table 3 shows this information and a brief description of the different modes.

Two mode shapes have been emphasized in this table due to their importance. The third experimental mode, corresponding to the natural frequency of 58.91 Hz, was identified in the machining tests as the one responsible of the most important roundness errors in the workpiece (figures 4, 5 and 6). Given its importance it is called the main chatter mode. Figure 8 shows a sequence of this mode, where one can appreciate that it is dominated by an important relative displacement between the grinding and the regulating wheel heads.

The ninth experimental mode selected in table 3, corresponding to a natural frequency of 129.86 Hz, was also excited during the machining tests. This mode affects mainly to the vibrations generated on the workblade and its influence in the roundness errors of the workpieces is less than the previous one. It is called the secondary chatter mode. In this mode, lower slide bending predominates, moving the workblade in its path. This is shown in figure 9.

Figures 8 and 9 show that when the main chatter mode is excited large displacements are generated on both wheels, whereas in the case of the secondary chatter mode the most significative displacements occur in the workblade.

2.3. Comparison between Numerical and Experimental Natural Frequencies

Once the numerical and the experimental analysis are completed in the centerless grinding machine, two different models of the same physical system are available. In theory both models should provide the same modal parameters but, in practice, a comparison of the natural frequencies obtained both numerically and experimentally (tables 1 and 3) shows that there is no exact correspondence between them.

The discrepancies observed may be due to numerical and experimental errors. Nevertheless, possible experimental errors are usually ignored and parameters obtained experimentally are assumed to be correct [9, 10]. Thus, considering the differences in the natural frequencies, it is concluded that an updating process is required to adjust the results obtained numerically to those obtained experimentally.

2.4. FE Model Updating Process

Model updating can be defined as an adjustment of an existing numerical model using as reference data obtained experimentally. The result is a FE model which represents the dynamic characteristics of the structure accurately. There are two updating philosophies:

Direct correction methods, where individual entries of system matrices are directly adjusted by comparison between initial numerical predictions and experimental results.

Indirect correction methods, where changes are made to specific physical properties of the numerical model.

The first philosophy generally leads to non-iterative methods, where the changes are not physically meaningful, i.e., they consist of assigning new values to individual entries of the matrices [M] and [K]. Corrections do not correspond to specific modelling errors, and the updated FE model is merely a matrix model whose dynamic behaviour is similar to the experimental model. Given these limitations, the updating process was performed using the second philosophy, which leads to methods that correct physical properties of the FE model. These methods are based on iterative procedures where two different phases are applied, a first one of error localization and a second one of error correction. It is assumed that the numerical model has a discrete number of errors which cause the discrepancies observed between predictions and measurements. Figure 10 shows the steps involved in an updating process for the chosen philosophy [11].

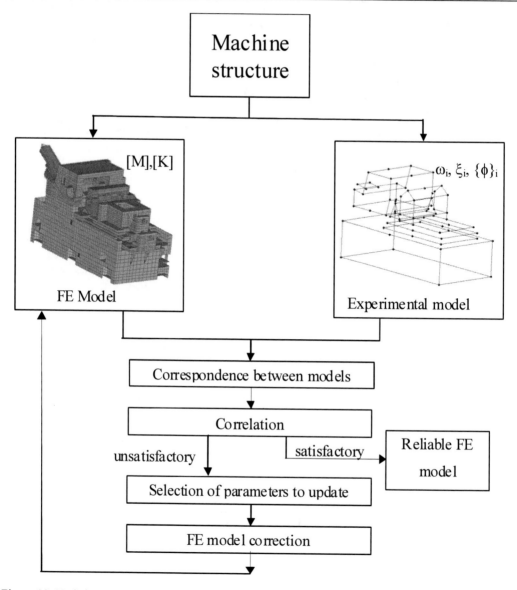

Figure 10. Updating process.

From the centerless grinding machine structure, the FE model and the experimental model are available. The FE model is characterised by the mass matrix [M] and the stiffness matrix [K]. As for damping, although causes producing it can be studied and integrated in the FE model [**Error! Reference source not found., Error! Reference source not found., Error! Reference source not found.**], it was ignored in this design phase due to the large amount of (non-linear) mechanisms that produce it, making it very difficult to develop a reliable model of its behaviour [15]. Damping properties of the different modes were considered in a subsequent phase, taking into account the modal dampings obtained experimentally [16].

The natural frequencies ω_i, modal dampings ξ_i and mode shapes $\{\phi\}_i$ were available from the EMA of the machine (table 3). These modal parameters were used to correct the FE model following the next sequence: correspondence between models, correlation, selection of parameters to update and FE model correction. These steps are explained in the following sections.

2.4.1. Correspondence between Models

The number of dof's of the FE model is much higher than the measured dof's. This is because the FE model requires a fine meshing in order to predict its dynamic behaviour, while it is unpractical to perform the experimental measurements in all these dof's. Furthermore, the experimental measurement points do not coincide exactly with the nodes of the FE model.

To find correspondence between the numerical and experimental dof's, the nodes of the FE model were compared in space with the points used in the EMA. This operation made it possible to define the nodes of the FE model which best corresponded to the points used in the EMA. The 69 points of the experimental database were paired with as many numerical nodes, defining a one-by-one correspondence between 207 dof's (3 dof's per node/point).

Table 4. Initial MAC values in % [Error! Reference source not found.]

FE mode shapes (Hz)	EMA mode shapes (Hz)									
	1 (33.41)	2 (48.43)	3 (58.91)	4 (76.97)	5 (90.52)	6 (108.44)	7 (122.02)	8 (128.48)	9 (129.86)	10 (144.02)
1 (30.16)	0.6	0.3	0.2	0.8	3.5	0.7	0.1	0.2	0.0	0.0
2 (31.84)	95.3	4.9	13.3	0.0	0.7	11.7	2.1	0.2	0.6	0.9
3 (45.9)	0.2	87.8	0.2	9.7	2.0	0.0	0.1	6.8	0.5	0.2
4 (54.85)	0.8	0.3	94.5	19.1	1.2	0.0	6.3	0.0	4.4	1.3
5 (67.38)	0.2	0.0	0.4	9.2	6.1	9.8	7.6	0.0	2.7	1.3
6 (73.5)	0.1	0.0	0.1	16.8	1.0	8.8	0.0	0.2	0.7	0.0
7 (82.27)	3.0	0.9	0.4	37.3	21.0	0.0	0.0	0.7	1.4	2.1
8 (93.03)	0.3	0.3	0.0	0.4	19.6	7.6	10.5	0.5	1.0	5.8
9 (96.84)	0.5	0.3	0.5	4.6	25.1	42.3	42.0	8.3	4.4	7.9
10 (102.02)	0.8	0.0	0.2	0.9	8.3	50.5	13.5	0.5	36.4	0.1
11 (113.28)	0.0	2.2	0.0	0.0	2.4	0.5	0.1	0.5	0.4	0.3
12 (122.13)	1.0	0.2	4.4	2.0	0.9	1.1	11.1	0.7	17.4	8.3
13 (126.82)	0.0	0.0	2.4	0.8	0.4	0.0	3.9	51.9	20.4	8.5
14 (149.88)	4.4	0.6	5.5	0.1	0.2	3.8	1.3	0.0	37.8	2.3
15 (159.44)	1.8	0.4	4.6	11.2	18.7	3.6	11.5	4.6	9.0	1.1

Table 5. Initial numerical-experimental pairing

Pair	FE mode	Hz	EMA mode	Hz	Difference (%)	MAC (%)
1	2	31.84	1	33.41	-4.70	95.3
2	3	45.9	2	48.43	-5.23	87.8
3	4	54.85	3	58.91	-6.90	94.5
4	10	102.02	9	129.86	-21.44	36.4

2.4.2. Correlation

The correlation techniques comprise a series of methods to compare both quantitatively and qualitatively the correspondence and difference between the modal parameters obtained numerically and experimentally.

Comparison between the numerical and experimental natural frequencies shown in tables 1 and 3 show that it is not easy to define a pairing between the experimental and the numerical modes. A widely used technique to express quantitatively the degree of correlation between mode shapes is the Modal Assurance Criterion, MAC [17]. The MAC value between a numerical mode, $\{\phi\}_{FE}$, and an experimental one, $\{\phi\}_{EMA}$, is defined as:

$$\text{MAC(FE, EMA)} = \frac{\left(\{\phi_{FE}\}^T \cdot \{\phi_{EMA}\}\right)^2}{\left(\{\phi_{FE}\}^T \cdot \{\phi_{FE}\}\right) \cdot \left(\{\phi_{EMA}\}^T \cdot \{\phi_{EMA}\}\right)} \qquad (14)$$

MAC values are always between 0 and 1, or between 0 % and 100 %. A MAC value equal to 100 % indicates an exact correlation between the two compared modes, while a MAC value equal to 0 % indicates that the two modes show no correlation at all.

Table 4 shows the MAC values obtained comparing the 15 numerical modes to the 10 experimental modes of the centerless grinding machine.

There are three MAC values highlighted in table 4, corresponding to as many mode pairs presenting very good correlation. Among them, it is interesting to emphasize the pair formed by the fourth numerical and the third experimental modes, with a MAC value of 94.5 %,. This is the main chatter mode, so it is concluded that the FE model is able to predict it accurately.

However, there are two problems requiring correction:

The three modes with good correlation present appreciable differences between the numerical and experimental natural frequencies.

No numerical mode presents an acceptable correlation with the secondary chatter mode, i.e., the ninth experimental mode at 129.86 Hz.

Therefore, it is necessary to improve the FE model via an updating process. In this phase the numerical modes to be corrected must be paired with the corresponding experimental modes. Pairing of the three modes highlighted in table 4 is straightforward, but there are difficulties when pairing the 9th experimental model. There are two numerical modes presenting MAC values close to 35 % with this experimental mode, i.e., the numerical mode n° 10 with a frequency of 102.02 Hz and the numerical mode n° 14 with a frequency of 149.88 Hz. It was decided to use the 10th numerical natural frequency for the pairing, as this natural frequency is lower than the experimental one to which it is to be approximated, presenting the same characteristic the three numerical frequencies paired initially. Table 5 shows this pairing.

The aim of the updating process is to modify some parameters of the FE model to reduce the differences between the numerical and experimental frequencies shown in table 5. At the same time, good correlation must be kept between the first three mode shape pairs and the correlation of the fourth pair must be improved substantially.

2.4.3. Selection of Updating Parameters

In this phase the parameters of the FE model that are not accurate and require correction must be identified. An adequate selection of these parameters is a very important task, although very difficult at the same time, since the parameters of the FE model causing bad correlation with the experimental data are not usually known.

The most suitable parameters to be modified are those with the greatest influence in the system responses (natural frequencies, mode shapes, etc.) and their correct selection is essential to guarantee the convergence of the FE model results towards the experimental data. Three groups of parameters were considered as candidates to be updated: the stiffness values of the joint elements between the different components, the lumped masses modelling the different motors of the machine and the densities of the elements modelling the two wheels.

Selected Stiffness Values

When a mechanical system is to be modelled, one of the major difficulties is the correct idealisation of the joints between the components [18, 19]. In the case of the centerless grinding machine, there are rigid joints (e.g. to model screwed joints) and joints that must allow a relative movement between the components (e.g. to model guidance systems). These joints were modelled using elastic elements providing stiffness in different directions, being the numerical values of these parameters very difficult to quantify and as such, with great uncertainty.

With the idea of limiting the number of stiffness values to update, a sensitivity analysis was performed using all the possible stiffnesses as parameters and considering the four numerical frequencies shown in table 5 as responses to be improved. This procedure showed that the stiffness values most affecting the estimation of the selected frequencies are the ones corresponding to the joints between the bed and the floor and the one corresponding to the lower spindle.

Selected Masses

The motors that guide the different spindles and the regulating wheel dresser were modelled as lumped masses joined stiffly to the machine structure. The mass values of these elements are parameters that present little uncertainty in the modelling process, but they idealize elements with relative movements in the structure, thereby affecting the inertia they provide.

Selected Densities

The grinding and the regulating wheels were modelled using several variable section beam elements. The density used to model these beam elements covered the density of the wheel material and the density of the wheel shaft material as a whole, being the exact determination of the numerical value of this resultant density a difficult task.

2.4.4. FE Model Correction

The last step in the updating process is the correction of the FE model modifying iteratively the selected parameters. On each iteration, the values of the updated parameters were obtained using a Bayesian parameter estimation technique [20, 21].

Table 6. MAC values after corrections, in %

FE mode shapes (Hz)	EMA mode shapes (Hz)									
	1 (33.41)	2 (48.43)	3 (58.91)	4 (76.97)	5 (90.52)	6 (108.44)	7 (122.02)	8 (128.48)	9 (129.86)	10 (144.02)
1 (32.53)	9	1.7	0.9	0.7	2.9	0.2	0.4	1.1	0.2	1.8
2 (33.45)	<u>90.1</u>	0.2	12.2	0.3	0	11.4	1.2	0.4	1	5.7
3 (48.22)	7.2	<u>91.8</u>	1.6	9.2	0.3	1.6	0.2	5.7	1.8	0
4 (58.91)	14	0	<u>96.4</u>	15.8	0.4	3	0.6	0.1	5.1	0.2
5 (74.94)	0.3	0	4.3	0.7	4.3	11.4	9.5	0	0	5.2
6 (86.03)	5.6	0	1.4	65.3	6.2	3.6	2	0	0	7.6
7 (90.22)	0	2.9	0.3	10.1	5.2	4.6	0.1	1.1	1	0.3
8 (98.31)	0	0.1	1	1.2	44.4	5.1	1.4	0.3	0.3	4.1
9 (104.34)	0.3	1	3.1	8.6	36.9	26.8	61.4	19.3	3	0.8
10 (116.23)	0	0.4	6.3	0.1	7	64.3	46.9	0.7	20.2	7.6
11 (118.32)	0.1	0.2	1.1	0	1.9	65.6	33	1.3	37	3.2
12 (125.66)	0.1	0.4	6.2	2.8	0.5	5.2	3	0.7	<u>75.4</u>	5.3
13 (139.52)	8	0.9	7	0.2	1.9	2.8	2.1	1.6	37.2	13.8
14 (145.61)	0.2	0	0	0	0	1	0	44.1	5.5	15.5
15 (162.52)	0.7	0	6.3	14.8	23.2	4.9	15.9	21.7	13.9	25.8

Table 7. Most significative results after updating

Pair	FE mode	Hz	AME mode	Hz	Difference (%)	MAC (%)
1	2	33.45	1	33.41	0.12	90.1
2	3	48.22	2	48.43	-0.44	91.8
3	4	58.91	3	58.91	-0.01	96.4
4	12	125.66	9	129.86	-3.24	75.4

After completing the updating process, a FE model with corrected parameter values was obtained. The updated MAC matrix is shown in table 6.

Table 7 shows a comparison between the natural frequencies of the updated numerical model and the frequencies obtained experimentally for the four paired modes, where the new MAC values can be appreciated.

Comparing this table with table 5, two important improvements are observed:

The difference between numerical and experimental natural frequencies was reduced for the first three pairs, maintaining their MAC values over 90 %.

MAC value for the fourth pair was increased significatively and, in addition, the difference between the numerical and the experimental natural frequency was reduced.

2.5. Study of the Updated FE Model

Once the updated FE model was obtained, a series of verifications were performed to validate the success of the updating process. These verifications consisted of checking that the FE model predicts the excitation of the main and the secondary chatter modes when the normal grinding force is applied and in comparing several numerically obtained FRFs with their respective experimental FRFs.

2.5.1. Study of the Most Important Modes

To identify the modes which are mostly excited under operating conditions, it is necessary to take into account that the normal force at the cutting point between the workpiece and the grinding wheel is produced mainly in Z direction. Therefore, the relative excitability of the different modes in this direction was evaluated calculating their modal participation factors (MPFs). With this purpose, the excitation was applied in Z direction in a node of the FE model located on the grinding wheel shaft, as this dof was considered as representative of the zone where the normal grinding force occurs. The result is shown in figure 11, where the MPFs have been normalized so the largest value has unit magnitude.

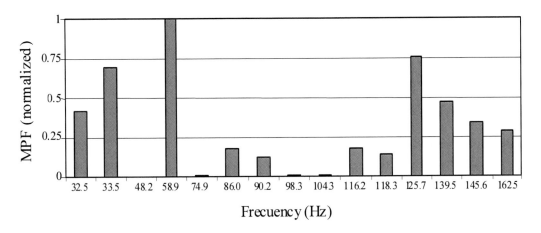

Figure 11. Modal participation factors.

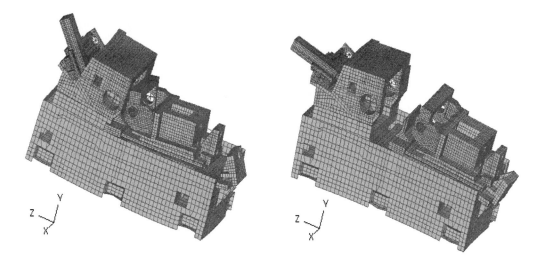

Figure 12. Main chatter mode [8].

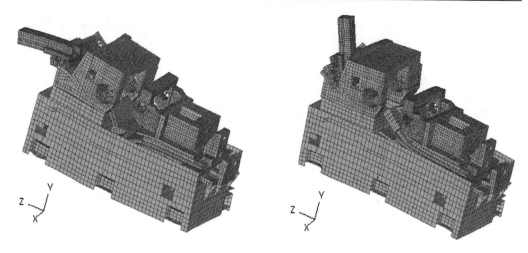

Figure 13. Secondary chatter mode.

In this figure, it can be seen that the FE model predicts a great excitation of the main chatter mode, with a natural frequency of 58.91 Hz, and the secondary chatter mode, with a natural frequency of 125.66 Hz. This result coincides with the experimental observations, indicating a correct qualitative prediction of the FE model in this aspect. Furthermore, there is also an important excitation of the mode at 33.45 Hz. Precisely, these three mode shapes and their respective natural frequencies were selected in the FE model updating process as responses to be improved.

In the mode at 33.45 Hz, all machine components move in phase in the longitudinal direction in a suspension movement with respect to the supports of the bed with the foundation.

The main chatter mode at 58.91 Hz is shown in figure 12.

This sequence shows clearly that the main mode is dominated by the out of phase movement of the two wheel heads, presenting a large relative displacement of the lower slide with respect to the swivel plate in Z direction. Comparing this figure with figure 8, where the same experimental mode was shown, it is seen that the numerical model predicts adequately this mode shape.

Finally, the secondary chatter mode at 125.66 Hz is shown in figure 13.

In this sequence, it can be seen that the secondary mode is dominated by the lower slide bending movement, moving the workblade in its path. There is a great similarity between this mode and its experimental equivalent in figure 9.

2.5.2. Comparison of FRFs

The updated FE model can be used to obtain different types of numerical FRFs (receptance, mobility, accelerance) between the different dof's. In practice, a widely used method to verify the precision of numerical models is to compare the estimated FRFs with the corresponding experimental ones.

To obtain the numerical FRFs, the damping properties must be included in the model. A modal damping approach was selected, where the damping properties obtained experimentally were used in the numerical mode shapes.

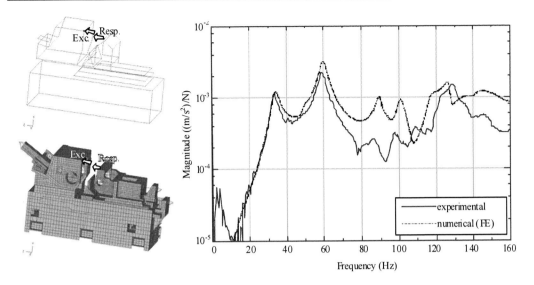

Figure 14. Numerical and experimental accelerances. Response point on the regulating wheel head.

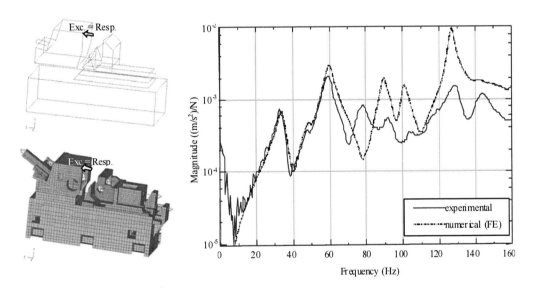

Figure 15. Numerical and experimental accelerances. Response point on the grinding wheel head.

Accelerances (acceleration/force) were selected as comparison FRFs because a series of 207 experimental response funtions of this type were available from the EMA. These response functions were obtained exciting the structure longitudinally in a point located on the grinding wheel head (figure 7), so the excitation dof selected in the numerical model was the same.

The responses were obtained in three points where large displacements were expected in chatter conditions.

The first response was obtained in the longitudinal direction of the machine at a point located on the regulating wheel head. The excitation and response dof's are shown in figure 14, together with the numerical and the experimental FRFs.

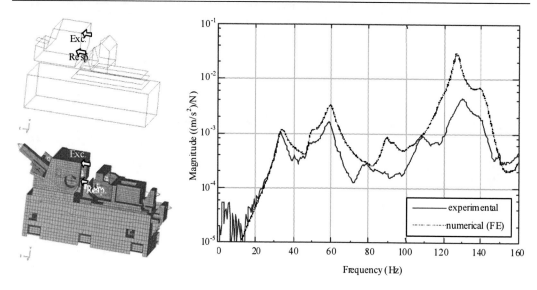

Figura 16. Numerical and experimental accelerances. Response point on the workblade.

This figure shows that the numerical model reflects correctly the dynamic properties of the system approximately below 70 Hz, whereas the discrepancies with the experimental results are greater above this frequency. Both the responses are dominated by the main chatter mode (approximately at 59 Hz). Furthermore, there are two more modes in the two curves with high contribution in the response: the secondary chatter mode at about 126 – 129 Hz and an additional mode at around 33 Hz. These three modes were identified in section 0 as the ones which were mostly excited in operating conditions.

Subsequently, the response was also obtained in the longitudinal direction but in this case at the same point of the grinding wheel head that had been used as excitation. Figure 15 shows the excitation and response dof's, as well as the FRFs obtained both numerically and experimentally.

In this case it is also possible to appreciate the similarities between the two responses at low frequencies, while at higher frequencies the discrepancies are clear. Again, the great contribution in the response of the same three modes shown in figure 14 can be observed.

Finally the response was obtained at a point located on the worblade, as shown in figure 16.

At low frequencies, the same considerations can be made as in the two previous comparisons. In this case, the response is dominated by the secondary chatter mode and, again, the important contribution in the response of the same three modes as in the previous cases can be observed.

3. DEVELOPMENT OF THE REDUCED MODEL

In section 2 a detailed dynamic modelling of the centerless grinding machine has been carried out, according to an updating process on the basis of data obtained experimentally. There is also evidence that the resultant FE model describes adequately the dynamic behaviour of the grinding machine in the feed direction of the upper slide. Instead, one of the biggest drawbacks of the model is the high number of dof's that it possesses. The use of this model in subsequent analysis would mean a huge computational cost, plus extensive simulation time. It is therefore necessary to reduce the order of the FE model so that the reduced model has the same frequency response characteristics as the original model in the frequency range of interest.

3.1. Damping Consideration

After the updating process, there is an improved FE model and its dynamic behaviour can be obtained from equation (1). This equation can be written as follows:

$$[M]\{\ddot{z}(t)\} + [K]\{z(t)\} = [L_u]\{f(t)\} \tag{15}$$

where $[L_u]$ is the input force influence matrix [22], indicating the spatial distribution of the input forces on the structure.

Equation (15) does not take into account the effects of damping in the structure, which were omitted in the FE model. However, due to the motion of the structure, dissipative forces are generated to be added to the inertial and elastic forces indicated in equation (15). The formulation of a mathematical expression for the dissipation of energy is complicated because, among the underlying causes of this dissipation, effects such as dry and viscous friction, hysteretic-type internal friction, damping elements, etc should be taken into consideration [15]. Moreover, in practice, the distribution of damping in the structure is not known with sufficient detail to warrant complicated modelling of the mentioned effects, so traditionally simplified models have been used. Among these models, one of the most widely accepted is the viscous damping formulation, which is to assume that the dissipative forces are proportional to the velocities. By adopting this solution, the dynamic equation of motion is:

$$[M] \cdot \{\ddot{z}(t)\} + [C] \cdot \{\dot{z}(t)\} + [K] \cdot \{z(t)\} = [L_u] \cdot \{f(t)\} \tag{16}$$

In this equation, [C] is the damping matrix of the system and $\{\dot{z}(t)\}$ is the vector of generalized velocities.

3.2. Obtaining State-Space Models

For the purpose of carrying out structural dynamic simulations on the centerless grinding machine, a convenient way to describe its behaviour is representing the system equations in state space form. In the state space form, a time invariant linear system is described by a series of first order differential equations:

$$\{\dot{x}(t)\} = [A] \cdot \{x(t)\} + [B] \cdot \{u(t)\} \tag{17}$$

$$\{y(t)\} = [C] \cdot \{x(t)\} + [D] \cdot \{u(t)\} \quad (18)$$

where [A], [B], [C] and [D] are real matrices independent of time. The variable t is time and the vector $\{u(t)\}$ is the system input. The vector $\{x(t)\}$ is the state vector and is composed of the state variables, while the vector $\{y(t)\}$ is the output of the system. The derivative term $\{\dot{x}(t)\}$ is the vector formed by the derivatives of each scalar term of $\{x(t)\}$. Equation (17) provides a complete description of the internal dynamics of the system and is called the state equation. Equation (18) is the output equation.

The matrices [A], [B], [C] and [D] depend on the choice made for the state vector; hence several state space realizations can be developed for the same inputs and outputs of the system.

3.2.1. Modal Coordinates

The orthogonality conditions referred to in section 0 can be used to make a change of variables in equation (16) from physical coordinates $\{z(t)\}$ to modal coordinates $\{p(t)\}$ as follows:

$$\{z(t)\} = [\Phi] \cdot \{p(t)\} \quad (19)$$

Because the matrix [Φ] is time invariant, the time derivatives of equation (19) are:

$$\{\dot{z}(t)\} = [\Phi] \cdot \{\dot{p}(t)\} \quad (20)$$

$$\{\ddot{z}(t)\} = [\Phi] \cdot \{\ddot{p}(t)\} \quad (21)$$

Substituting these coordinate changes in equation (16) the following is obtained:

$$[M] \cdot [\Phi] \cdot \{\ddot{p}(t)\} + [C] \cdot [\Phi] \cdot \{\dot{p}(t)\} + [K] \cdot [\Phi] \cdot \{p(t)\} = [L_u] \cdot \{f(t)\} \quad (22)$$

Premultiplying by $[\Phi]^T$ and using the orthogonality conditions given in (12) and (13) provides:

$$[I] \cdot \{\ddot{p}(t)\} + [\Phi]^T \cdot [C] \cdot [\Phi] \cdot \{\dot{p}(t)\} + [\Omega^2] \cdot \{p(t)\} = [\Phi]^T \cdot [L_u] \cdot \{f(t)\} \quad (23)$$

At this point it is necessary to study the damping matrix [C] present in equation (23). Due to the complication of establishing an analytical procedure to determine its value, it is common to use a simplified procedure for calculating damping without having to resort to detailed modelling based on structural properties of the materials used.

Two simplified procedures are popular in practice, proportional damping (Rayleigh) and modal damping [10]. Between them, the second is particularly adapted to the use of the results obtained from the EMA, so this is what was used in this development. In this procedure, the magnitude of the damping for each mode is expressed as a percentage of critical damping, and it gives adequate results provided that the damping of the different modes is weak, typically less than 10% of critical damping [10], as occurs in the centerless grinding machine.

Using the procedure mentioned, the matrix equation (23) is transformed in a series of N individual decoupled equations, each of which describes the behaviour of a mode shape. For each mode, it is possible to consider the modal fraction of critical damping ξ_i (modal damping), defined from:

$$[\Phi]^T \cdot [C] \cdot [\Phi] = [\text{diag}(2 \cdot \xi_i \cdot \omega_i)] \tag{24}$$

and equation (23) can be rewritten as:

$$\{\ddot{p}(t)\} + 2 \cdot [\xi] \cdot [\Omega] \cdot \{\dot{p}(t)\} + [\Omega^2] \cdot \{p(t)\} = [\Phi]^T \cdot [L_u] \cdot \{f(t)\} \tag{25}$$

where $[\xi]$ represents the diagonal matrix whose terms are the damping ratios for each mode, i.e.:

$$[\xi] = \begin{bmatrix} \xi_1 & 0 & \cdots & 0 \\ 0 & \xi_2 & \cdots & 0 \\ \cdots & \cdots & \cdots & \cdots \\ 0 & 0 & \cdots & \xi_N \end{bmatrix} \tag{26}$$

In equation (25) it is only necessary to calculate the damping ratios ξ_i and therefore the damping matrix [C] is not obtained explicitly. This represents an advantage, because the values of ξ_i for different modes can be obtained experimentally.

The $[\xi]$ and $[\Omega]$ matrices have a diagonal shape, so that the N decoupled equations obtained from (25) can be written as:

$$\ddot{p}_i(t) + 2 \cdot \xi_i \cdot \omega_i \cdot \dot{p}_i(t) + \omega_i^2 \cdot p_i(t) = \{\phi\}_i^T \cdot [L_u] \cdot \{f(t)\}$$
$$i = 1, 2, \ldots, N \tag{27}$$

where $\{\phi\}_i^T \cdot [L_u]$ indicates row i of the matrix $[\Phi]^T \cdot [L_u]$. Each equation (27) defines the damped motion of a system with one dof.

3.3. Reduction Procedure

Modal equation (25) has the same order as the equation in physical coordinates (16), so the change of coordinates (19) does not show the advantages gained in this area by using modal coordinates. These advantages come from the fact that it is possible to present the equation (25) considering only the modal parameters corresponding to the first 15 mode shapes without calculating the rest, as their dynamic contribution in the frequency range of interest is neglected.

To this end, the matrix of mode shapes [Φ] is decomposed in two parts as follows:

$$[\Phi] = [[\Phi]_m \quad [\Phi]_{N-m}] \tag{28}$$

where $[\Phi]_m$ contains the first m = 15 calculated mode shapes, whereas $[\Phi]_{N-m}$ contains the remaining modes up to the number of dof's of the structure, which have not been calculated. Using this approach, it is possible to apply the orthogonality conditions given by (12) and (13) only for the known modes:

$$[\Phi]_m^T \cdot [K] \cdot [\Phi]_m = [\Omega^2]_m \tag{29}$$

$$[\Phi]_m^T \cdot [M] \cdot [\Phi]_m = [I]_m \tag{30}$$

where $[\Omega^2]_m$ is the diagonal matrix containing the squares of the first 15 natural frequencies and $[I]_m$ is the identity matrix of dimension 15 × 15.

Taking into account the division (28), the change of variables given by equation (19) may be written as follows:

$$\{z(t)\} = [[\Phi]_m \quad [\Phi]_{N-m}] \cdot \begin{Bmatrix} \{p(t)\}_m \\ \{p(t)\}_{N-m} \end{Bmatrix} \tag{31}$$

where $\{p(t)\}_m$ is the vector containing the modal coordinates corresponding to the calculated modes, while $\{p(t)\}_{N-m}$ contains those corresponding to non-calculated modes. Equation (31) can be written as:

$$\{z(t)\} = [\Phi]_m \cdot \{p(t)\}_m + [\Phi]_{N-m} \cdot \{p(t)\}_{N-m} = \{z(t)\}_m + \{z(t)\}_{N-m} \tag{32}$$

The most direct procedure to reduce the order of equation (16) is to remove or truncate from equation (32) the part corresponding to the non-calculated modes, leading to the modal truncation method. In this case, the vector of generalized displacement is expressed as:

$$\{z(t)\} \approx \{z(t)\}_m = [\Phi]_m \cdot \{p(t)\}_m \tag{33}$$

The derivatives of equation (33) are:

$$\{\dot{z}(t)\} \approx \{\dot{z}(t)\}_m = [\Phi]_m \cdot \{\dot{p}(t)\}_m \qquad (34)$$

$$\{\ddot{z}(t)\} \approx \{\ddot{z}(t)\}_m = [\Phi]_m \cdot \{\ddot{p}(t)\}_m \qquad (35)$$

Using these changes of coordinates the following is finally obtained:

$$\{\ddot{p}(t)\}_m + 2 \cdot [\xi]_m \cdot [\Omega]_m \cdot \{\dot{p}(t)\}_m + [\Omega^2]_m \cdot \{p(t)\}_m = [\Phi]_m^T \cdot [L_u] \cdot \{f(t)\} \qquad (36)$$

Equation (36) includes the modal parameters corresponding to the first 15 calculated mode shapes without taking into account the effect of the rest. However, the non-calculated modes may have a considerable contribution in the structural behaviour. To ensure that the dynamic responses obtained through the solution are sufficiently accurate, all the modes whose frequencies extend to at least two or three times the greater excitation frequency must be retained [23]. This requirement would mean making the calculation of several additional modes to the 15 already calculated, which would imply considering in the analysis various modes outside the range of interest, increasing the order of the resulting model with the resulting increase in computational cost. This drawback would eliminate the computational benefits obtained by modal truncation.

The disadvantage of the slow convergence of the modal truncation method can be solved by using the concept of residual flexibility, leading to the modal acceleration method [23], described in the next section.

3.3.1. Modal Acceleration Method

The modal acceleration method is based on correcting the response obtained by the modal truncation method assuming that the non-calculated modes contribute to the response in a quasistatic manner. Consequently, the response due to the non-calculated modes has no dynamic amplification, not producing significant velocities and accelerations. Thus, equation (32) is approximated by:

$$\{z(t)\} \approx \{z(t)\}_m + \{z(t)\}_{N-m}^{quasi} \qquad (37)$$

The first term on the right of equation (37) represents the response obtained using modal truncation, while the second is the corrector term which takes into account the static contribution of higher modes, which have not been calculated. According to equation (32) the corrector term $\{z(t)\}_{N-m}^{quasi}$ can be obtained as:

$$\{z(t)\}_{N-m}^{quasi} = \{z(t)\}^{quasi} - \{z(t)\}_m^{quasi} \qquad (38)$$

In this expression, $\{z(t)\}^{quasi}$ is obtained as a solution of equation (16) establishing $\{\dot{z}(t)\}^{quasi} = \{\ddot{z}(t)\}^{quasi} = \{0\}$, leading to:

$$[K] \cdot \{z(t)\}^{quasi} = [L_u] \cdot \{f(t)\} \qquad (39)$$

which solution is:

$$\{z(t)\}^{quasi} = [K]^{-1} \cdot [L_u] \cdot \{f(t)\} \qquad (40)$$

The term $\{z(t)\}_m^{quasi}$ is obtained, based on equality (33), as:

$$\{z(t)\}_m^{quasi} = [\Phi]_m \cdot \{p(t)\}_m^{quasi} \qquad (41)$$

where $\{p(t)\}_m^{quasi}$ represents the solution of equation (36) when $\{\dot{p}(t)\}_m = \{\ddot{p}(t)\}_m = \{0\}$, i.e.:

$$\{p(t)\}_m^{quasi} = [\Omega^{-2}]_m \cdot [\Phi]_m^T \cdot [L_u] \cdot \{f(t)\} \qquad (42)$$

therefore, according to equation (41), the following is obtained:

$$\{z(t)\}_m^{quasi} = [\Phi]_m \cdot [\Omega^{-2}]_m \cdot [\Phi]_m^T \cdot [L_u] \cdot \{f(t)\} \qquad (43)$$

Substituting equations (40) and (43) in (38) leads to:

$$\{z(t)\}_{N-m}^{quasi} = [K]^{-1} \cdot [L_u] \cdot \{f(t)\} - [\Phi]_m \cdot [\Omega^{-2}]_m \cdot [\Phi]_m^T \cdot [L_u] \cdot \{f(t)\} \qquad (44)$$

Finally, equation (37) is:

$$\{z(t)\} \approx [\Phi]_m \cdot \{p(t)\}_m + [K]^{-1} \cdot [L_u] \cdot \{f(t)\} - [\Phi]_m \cdot [\Omega^{-2}]_m \cdot [\Phi]_m^T \cdot [L_u] \cdot \{f(t)\} \qquad (45)$$

The first term on the right side of this equation is the same as that obtained in equation (33) applying the modal truncation method. The second and third terms represent the static correction of the truncated modes, and it is worth mentioning that they are obtained without having to calculate more than the first 15 modal parameters. With the introduction of these correction terms, the number of required modes is significantly less than those required by the modal truncation method to obtain the same precision in the response [23]. The condition of convergence of the modal acceleration method is that the lower frequency of the truncated modes must be greater than the upper limit of the excitation frequencies [23], so with the 15 calculated modes obtaining an adequate response is guaranteed.

For the derivative terms of $\{\dot{z}(t)\}$ the expressions (34) and remain valid because it was considered that the response due to the non-calculated modes does not contribute to the velocities and accelerations.

The additional cost of the modal acceleration method comes from the need to calculate the term $[K]^{-1}$, which indicates that it is necessary to perform a static analysis of the grinding machine based on the FE model, in addition to modal analysis already done. The practical procedure to obtain this term is explained at the end of this section.

Selected Dof's

The development of the modal acceleration method has been carried out considering all the dof's of the FE model. However, many of these dof's are of no practical interest, and the previous development can be done taking into account only those which are representative of the dynamic behaviour of the structure. Thus, to construct the reduced model only the dof's corresponding to the X, Y and Z directions of the 69 nodes of the FE model previously paired with the measurement points have been selected, leading to 207 dof's.

Thus, the matrix of mode shapes $[\Phi]_m$ is constructed by extracting the modal displacements of the first 15 modes for the 207 selected dof's. Similarly, $[L_u]$ contains unit terms in the locations corresponding to the dof's where the excitation force $\{f(t)\}$ acts, being null the rest of its terms until completing the 207 dof's.

As for the term $[K]^{-1}$, it must be noted that obtaining it by calculating the inverse of the global stiffness matrix of the FE model would be computationally unapproachable. In addition, the calculation of all the terms of this matrix is not necessary, since as it is apparent from the number of selected dof's, its dimensions must be 207 × 207. It was calculated based on equation, which has been defined as the solution of the static analysis carried out on the FE model. If this equation is applied only for the 207 selected dof's, the term i,j of matrix $[K]^{-1}$ refers to the displacement obtained in the i dof due to a unit force applied to the j dof. Therefore, this matrix was obtained applying sequentially a unit force on each of the 207 selected dof's and calculating for each force the displacements in these dof's

3.3.2. State Space Representation

After selecting the dof's to consider during the analysis, the modal parameters corresponding to the first 15 modes were grouped in the matrix of mode shapes $[\Phi]_m$, the matrix of natural frequencies $[\Omega]_m$ and matrix of damping ratios $[\xi]_m$. As mentioned above, the terms ξ_i of the latter matrix were determined from the experimentally obtained modal damping ratios. The four paired numerical modes in table 7 were assigned the modal damping ratios corresponding to the respective experimental modes. For the other modes of the FE model that were not correlated during the updating process a modal damping value equal to 1.5% was chosen, which is a typical damping value in structures similar to grinding machine under study [24].

It is noteworthy that the difference between equations (25) and (36) is that the first one has been developed for all the modes of the machine, while the second has been developed only for the first 15 modes. Therefore, equation (36) represents a series of uncoupled equations of the form:

$$\ddot{p}_i(t) + 2\cdot\xi_i\cdot\omega_i\cdot\dot{p}_i(t) + \omega_i^2\cdot p_i(t) = \{\phi\}_i^T\cdot[L_u]\cdot\{f(t)\} \qquad (46)$$
$$i = 1, 2...,15$$

Using these modal coordinates the state space representation using these modal coordinates can be defined selecting for each mode of equation (46) the following state variables [22, 25]:

$$\{x(t)\}_i = \begin{Bmatrix} \omega_i \cdot p_i(t) \\ \dot{p}_i(t) \end{Bmatrix} \qquad (47)$$

thus the global state vector is obtained as follows:

$$\{x(t)\} = \begin{Bmatrix} \{x(t)\}_1 \\ \{x(t)\}_2 \\ \dots \\ \{x(t)\}_{15} \end{Bmatrix} = \begin{Bmatrix} \omega_1 \cdot p_1(t) \\ \dot{p}_1(t) \\ \omega_2 \cdot p_2(t) \\ \dot{p}_2(t) \\ \dots \\ \omega_{15} \cdot p_{15}(t) \\ \dot{p}_{15}(t) \end{Bmatrix} \qquad (48)$$

This last expression shows that the size of the state vector is 30, i.e., twice the modes included in the model. Each mode has two associated states.

A convenient way to express the state vector (48) is grouping the terms corresponding to $\omega_i\cdot p_i(t)$ and $\dot{p}_i(t)$:

$$\{x(t)\} = \begin{Bmatrix} \omega_1 \cdot p_1(t) \\ \omega_2 \cdot p_2(t) \\ \dots \\ \omega_{15} \cdot p_{15}(t) \\ \dot{p}_1(t) \\ \dot{p}_2(t) \\ \dots \\ \dot{p}_{15}(t) \end{Bmatrix} \qquad (49)$$

which can be expressed as:

$$\{x(t)\} = \begin{bmatrix} [\Omega]_m \cdot \{p(t)\}_m \\ \{\dot{p}(t)\}_m \end{bmatrix} \quad (50)$$

Through this definition of the state vector and bearing in mind (36), the state equation is:

$$\{\dot{x}(t)\} = \begin{bmatrix} [0] & [\Omega]_m \\ -[\Omega]_m & -2\cdot[\xi]_m \cdot [\Omega]_m \end{bmatrix} \cdot \{x(t)\} + \begin{bmatrix} [0] \\ [\Phi]_m^T \cdot [L_u] \end{bmatrix} \cdot \{f(t)\} \quad (51)$$

The order of this equation is 30, considerably lower than the order of the initial FE model.

As for the output equation, its form depends on the required output, so that the model developed can be used to obtain displacement, velocity and acceleration of the dof's considered.

3.4. Comparison of FRFs

Once the reduced order model was obtained in state space form, its validity was checked by comparing several FRFs obtained using this reduced model with those obtained with the FE model. As was done in section 0, the excitation dof was selected in the grinding wheel head and the response dof's in three locations, the regulating wheel, the grinding wheel head and the workblade. Figures 17, 18 and 19 show the resulting accelerances.

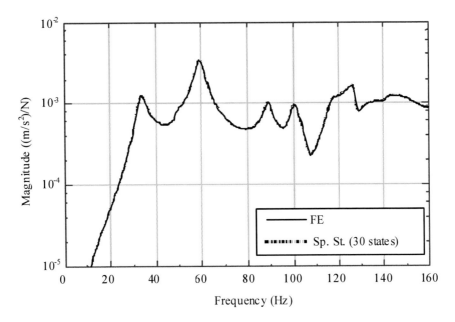

Figure 17. Comparisons of FRFs. Response point on the regulating wheel head.

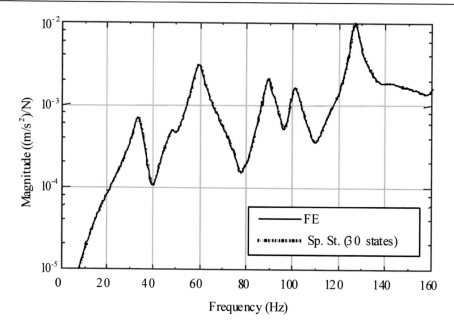

Figure 18. Comparisons of FRFs. Response point on the grinding wheel head.

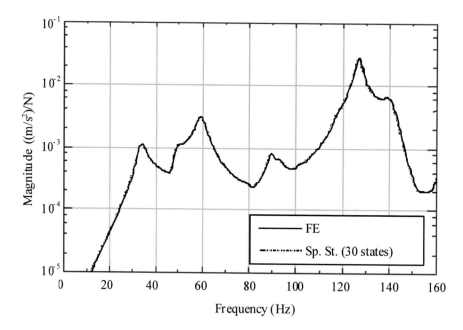

Figure 19. Comparisons of FRFs. Response point on the workblade

These figures show that the reduced state space model reflects almost exactly the dynamic behaviour of the structure estimated by the FE model, demonstrating the effectiveness of the reduction procedure.

4. VALIDATION OF THE REDUCED MODEL

In section 3 a reduced order state space model of the grinding machine has been obtained in order to study its vibration characteristics.

The next step in the development of this chapter is to validate this model, estimating theoretically different machine responses under chatter conditions, both in the frequency domain and in the time domain. The comparison between numerical predictions and experimental measurements make it possible to assess the accuracy of the simulation procedure.

4.1. Block diagram of the Process

The theoretical study of the dynamic instabilities associated with the regenerative chatter in the centerless grinding process was made starting from the block diagram of the process shown in figure 20, which has been obtained from similar diagrams developed previously in the literature [4, 5].

In this figure, the angles φ_1 and φ_2 (see figure 1) and the terms $\varepsilon' = \dfrac{\sin \varphi_2}{\sin(\varphi_2 - \varphi_1)}$ and $(1-\varepsilon) = \dfrac{\sin \varphi_1}{\sin(\varphi_2 - \varphi_1)}$ are variables depending on the geometrical configuration of the machine, s is the Laplace operator, ω_p is the worpiece angular velocity, K is the cutting stiffness and k'_{cg} and k'_{cr} are the specific contact stiffnesses between the grinding wheel and the workpiece and between the regulating wheel and the workpiece, respectively.

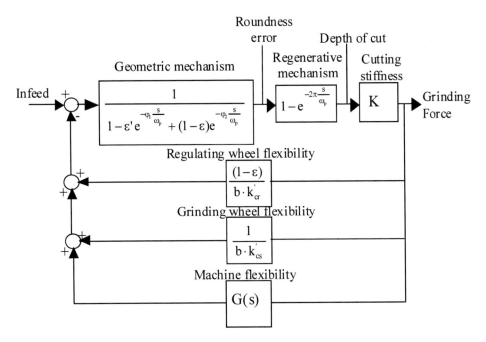

Figure 20. Block diagram of the centerless grinding process.

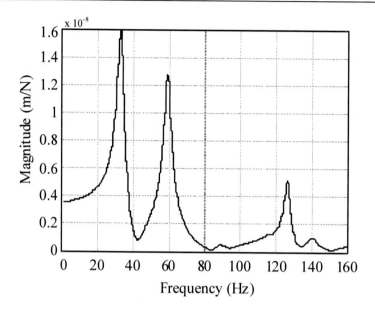

Figure 21. Dynamic flexibility of the machine, G(s).

The function G(s) represents the dynamic flexibility of the machine. Figure 21 shows the FRF corresponding to this function, which has been obtained using the state space model of order 30.

This curve shows that in the dynamic response it is only important the participation of the three modes that have been remarked in previous sections. The rest of the modes have little contribution. Therefore, the use of a transfer function of order 30 for this representation is unnecessary and also computationally expensive.

Therefore, it is interesting to develop an additional reduction procedure leading to a reduced representation considering only dynamic properties of the previous three modes. To develop this reduction procedure, it must be considered that the modes with highest contribution in a given FRF are those which are best excited by the input dof and/or those which present the greatest modal response in the output dof.

The above idea can be properly developed using the concepts of controllability and observability [26]. Both concepts can be used in the model reduction process defining as the uncontrollable modes those that are not excited by the applied excitation and as the unobservable modes those that are not measured in the response. Recalling that the definition of the state vector (50) implies that each mode has two associated states, an uncontrollable or unobservable mode leads to two associated states that do not contribute to the FRF between the excitation and the response. Therefore, these states can be removed from the state space representation without affecting the input-output behaviour.

The described reduction procedure is based on the calculation of controllability and observability gramians. These gramians are defined as follows [22, 27]:

Controllability gramian:

$$[W_c] = \int_0^\infty e^{[A]\tau} \cdot [B] \cdot [B]^T \cdot e^{[A]^T \cdot \tau} \cdot d\tau \qquad (52)$$

Observability gramian:

$$[W_o] = \int_0^\infty e^{[A]^T \cdot \tau} \cdot [C]^T \cdot [C] \cdot e^{[A] \cdot \tau} \cdot d\tau \qquad (53)$$

These gramians are, respectively, the solutions of the following Lyapunov equations:

$$[A] \cdot [W_c] + [W_c] \cdot [A]^T + [B] \cdot [B]^T = [0] \qquad (54)$$

$$[A]^T \cdot [W_o] + [W_o] \cdot [A] + [C]^T \cdot [C] = [0] \qquad (55)$$

The gramians are square matrices where the main diagonal terms indicate the controllability and observability of the different states. To carry out the elimination of the less observable and less controllable states, the model must be transformed to the balanced realization [26], where the different states are as controllable as they are observable. This is achieved by an appropriate linear transformation of the state variables:

$$\{\bar{x}(t)\} = [T] \cdot \{x(t)\} \qquad (56)$$

where [T] is the transformation matrix. Thus the equations (17) and (18) are transformed to equations (57) and (58):

$$\{\dot{\bar{x}}(t)\} = [T] \cdot [A] \cdot [T]^{-1} \cdot \{\bar{x}(t)\} + [T] \cdot [B] \cdot \{u(t)\} \qquad (57)$$

$$\{y(t)\} = [C] \cdot [T]^{-1} \cdot \{\bar{x}(t)\} + [D] \cdot \{u(t)\} \qquad (58)$$

In this balanced realization, the controllability and observability gramians are identical and strictly diagonal. The terms of the main diagonal are the Hankel singular values of the system, and provide information about the relative controllability and observability of the different states.

The Hankel singular values obtained from the balanced realization of the centerless grinding machine are shown in figure 22, which have been arranged in order of relevance.

This figure shows that there are six states that stand above the rest, so the rest of the states have been residualized in order to consider their static contribution in the response [28]. Figure 23 shows the comparison between the FRF corresponding to the state space model with 30 states (continuous curve) and the residualized one with 6 states (dotted line).

It can be seen that the three modes associated with the representation of 6 states are the same as those which provide the largest contribution to the response in figure 21. Moreover, the representation of 6 states reproduces faithfully the behaviour of the representation of 30 states, so it checks the validity of the reduction procedure.

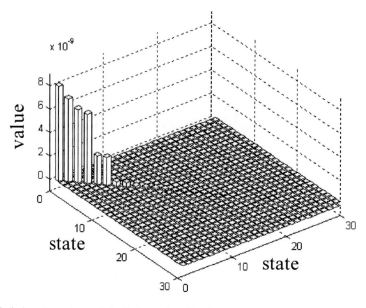

Figure 22. Hankel singular values of the balanced realization.

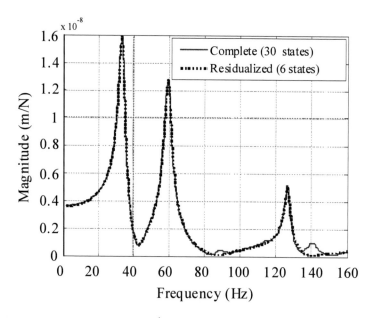

Figure 23. Receptance FRFs obtained with 30th order state space system and the 6th order system.

4.2. Simulations in the Frequency Domain

The characteristic equation of the block diagram shown in figure 20 is:

$$\left(1-e^{-2\pi\frac{s}{\omega_p}}\right)\cdot K\cdot\left[\frac{(1-\varepsilon)}{b\cdot k'_{cr}}+\frac{1}{b\cdot k'_{cg}}+G(s)\right]+1-\varepsilon'e^{-\varphi_1\frac{s}{\omega_p}}+(1-\varepsilon)e^{-\varphi_2\frac{s}{\omega_p}}=0 \quad (59)$$

The roots of this equation are usually complex, so they can be expressed by their real and imaginary part:

$$s = \sigma + j\omega \quad (60)$$

If all the characteristic roots are on the left side of the complex plane, the real parts of the roots are all negative and the system is stable. However, if at least one of the characteristic roots is located on the right side of the complex plane, the system is unstable and during the grinding process the structural response grows in time producing on the workpiece the regeneration of an appreciable roundness error. In this case, the real part of the unstable root, σ, indicates the growth rate of the surface ondulations generated in the workpiece (also called lobes) and the imaginary part indicates the frequency ω to which the machine vibrates. This frequency is related to the number of lobes generated in the part by the following expression:

$$n = \frac{\omega}{\omega_p} \quad (61)$$

4.2.1. Resolution Methodology

The complete resolution of the roots of the characteristic equation is not an easy task because of the nature of the transcendental equation to solve, consisting of three time delays, so there are infinite solutions that satisfy it.

In this chapter, the roots of the characteristic equation have been found using the root locus technique, which was introduced as a method to study the stability against chatter of the orthogonal cutting process by Olgac et al. [29]. To apply of the method it is taken into account that the equation (59) can be studied as the characteristic equation of the block diagram shown in figure 24.

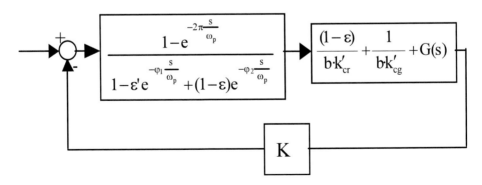

Figure 24. Modified block diagram.

In this figure, the term $\dfrac{1-e^{-2\pi\frac{s}{\omega_p}}}{1-\varepsilon' e^{-\varphi_1\frac{s}{\omega_p}}+(1-\varepsilon)e^{-\varphi_2\frac{s}{\omega_p}}} \cdot \left(\dfrac{(1-\varepsilon)}{b \cdot k'_{cr}}+\dfrac{1}{b \cdot k'_{cg}}+G(s)\right)$ is the open loop transfer function, while the cutting stiffness K is the feedback gain. The method of root locus is based on obtaining the solutions of the characteristic equation for increasing values of the cutting stiffness in the range $0 \to \infty$. Using this technique the evolution of the different roots is obtained, so it can be determined which one becomes unstable and for what value of the cutting stiffness.

The representation procedure starts solving the roots of the characteristic equation for K = 0, as described in [8]. Subsequently the first increase is provided in the value of K. Bearing in mind that for a small enough increase each root will be located very close to its original location, the solutions obtained with K = 0 are used as initial estimates of the solutions for this first value of K. Following an iterative process based on the Newton-Raphson method [30], the roots for this first value of K are obtained. The roots thus obtained are used as initial estimates for the next increment and, once found the roots for the second increase, the procedure continues until the cutting stiffness reaches the estimated value for the cutting conditions being considered.

4.2.2. Results

Figure 25 shows the evolution of the root locus for a particular configuration of the machine, with a workpiece angular velocity of ω_p = 8 Hz.

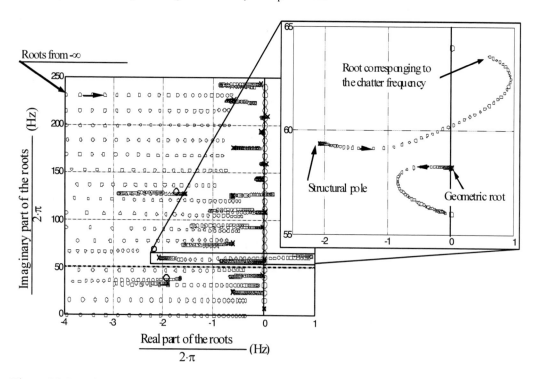

Figure 25. Root locus for increasing values of K.

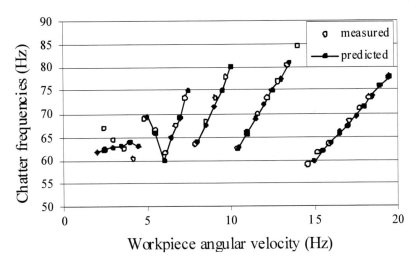

Figure 26. Comparison between theoretical and experimental chatter frequencies [8].

This figure shows that the structural pole at 58.91 Hz migrates towards the imaginary axis for increasing values of the cutting stiffness until the pole crosses it and, therefore, it is instabilized. The chatter frequency is determined by the imaginary part of the unstable root after completing the representation. In this case, the chatter frequency corresponds approximately to 64 Hz.

This procedure was repeated for values of the workpiece angular velocity in the 0.5-20 Hz range and chatter frequencies estimated theoretically were compared with the ones obtained experimentally under the same cutting conditions. Figure 26 shows this comparison.

This figure shows that the theoretical predictions are in good agreement with the experimental results.

4.3. Time Domain Simulations

In the previous section a stability analysis of the centerless grinding process has been undertaken, finding the roots of the characteristic equation (59). With this technique, qualitative information about the dynamic instability of the process can be obtained, in the sense that it is possible to determine whether a particular configuration is stable or not. However, the major drawback of the used method is the fact that it only allows linear phenomena to be considered. It is therefore not possible to take into account the nonlinear effects occurring when machining workpieces, such as the consideration of an arbitrary initial workpiece profile, the loss of contact between grinding wheel and workpiece, the spark-out process, etc... Because of this limitation, the frequency domain method developed previously cannot be used to quantitatively assess the severity of the instabilities. Even if the real parts of the obtained unstable roots can be regarded as quantitative measures of the growth rate of the workpiece lobes, this growth is limited by the mentioned nonlinearities.

To overcome the above limitation, it is very interesting to simulate the evolution of the centerless grinding process in time domain. This technique allows realistic simulations of the process through a quantitative evaluation of the surface finish of the machined workpieces.

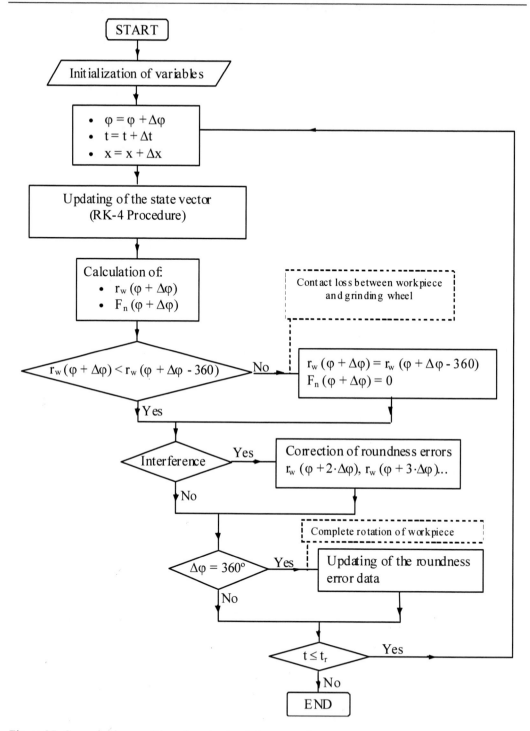

Figure 27. General scheme of time domain simulation procedure.

4.3.1. Mathematical Basis of the Developed Model

For the development of time domain simulations, initially the workpiece was discretized in 360 equal segments. Starting from the block diagram shown in figure 20, the time evolution of the workpiece roundness error was obtained a a function of the dynamic properties of the machine, the feed rate and the roundness errors of the workpiece part on the previous pass and on the points of contact with the workblade and the regulating wheel. Workpiece rotation was simulated segment by segment, updating at each step the roundness error generated by integrating its time evolution using the Runge-Kutta algorithm [30]. In this process, nonlinear effects of the process were taken into account, such as loss of contact between the workpiece and the grinding wheel, interference with the grinding wheel and spark-out process. The simulation process is shown in the diagram in figure 27.

4.3.2. Results

Time domain simulation procedure described in previous section was used to obtain the theoretical profile of the workpiece under certain cutting conditions. Figure 28a shows the theoretical profile, while figure 28b shows the real profile obtained experimentally for the same conditions.

It can be seen that the two profiles are very similar, showing the same number of lobes. In addition, the theoretical roundness error is of the same order of magnitude of the experimentally measured error.

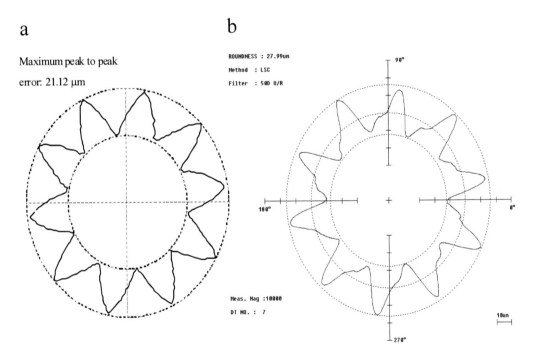

Figure 28. Final profile of the workpiece: (a) theoretical and (b) experimental [31].

5. CONCLUSION

In this chapter a comprehensive methodology for the theoretical study of the dynamic properties of a centerless grinding machine has been developed. The main conclusions are the following:

The updating process of the FE model of the machine was completed successfully. The objective of this process was not to seek an exact correlation between all the numerical and experimental mode shapes, but to improve the results of the modes that have a great influence in the dynamic characteristics of the machine.

The updated FE model was used to study the different modes, and it was determined that three of them contribute most to the vibrations of the machine (the main and secondary chatter modes and a third suspension mode). This result is consistent with the experimental measurements. In addition, the model predicts adequately the relative importance of these modes in the vibrations generated in different parts of the structure. Therefore, it is concluded that the model predicts correctly the dynamic behaviour of the machine.

The updated FE model has a high number of dof's, making it necessary to undergo a reduction process. A low order state space model was obtained, which adequately predicted the dynamic behaviour of the machine in the frequency range of interest.

The incorporation of the reduced model in the cutting model of the centerless grinding process made it possible to predict different responses of the machine, showing a very good correlation with experimental results. Therefore, a powerful tool has been developed with which the optimum operating conditions of the machine can be defined, enabling increased productivity.

ACKNOWLEDGMENTS

The authors are grateful for the subsidies granted by the University of the Basque Country (EHU08/44-2008).

REFERENCES

[1] Garitaonandia, I., Fernandes, M.H., Albizuri, J., Hernández, J.M. and Barrenetxea, D. (2010) A new perspective on the stability study of centerless grinding process, *International Journal of Machine Tools and Manufacture, 50*, 2, 165-173. With permission from Elsevier to reprint figures.

[2] Gurney, J.P. (1964) An analysis of centerless grinding, *Journal of Engineering for Industry, 87*, 163-174.

[3] Rowe, W.B. and Koenigsberger, F. (1965) The "work-regenerative" effect in centerless grinding, *International Journal of Machine Tool Design and Research, 4*, 3, 175-187.

[4] Furukawa, Y., Miyashita, M. and Shiozaki, S. (1971) Vibration analysis and work-rounding mechanism in centerless grinding, *International Journal of Machine Tool Design and Research, 11*, 2, 145-175.

[5] Miyashita, M., Hashimoto, F. and Kanai, A. (1982) Diagram for selecting chatter free conditions of centerless grinding, *Annals of the CIRP, 31*, 1, 221-223.

[6] R. Bueno, R., Zatarain, M. J. and Aguinagalde, M. (1990) Geometric and dynamic stability in centerless grinding, *Annals of the CIRP, 39*, 1, 395-398.

[7] Nieto, F.J., Etxabe, J.M. and Gimenez, J.G. (1998) Influence of contact loss between workpiece and grinding wheel on the roundness errors in centreless grinding, *International Journal of Machine Tools and Manufacture, 38*, 10-11, 1371-1398.

[8] Garitaonandia, I., Fernandes, M.H. and Albizuri, J. (2008) Dynamic model of a centerless grinding machine based on an updated FE model, *International Journal of Machine Tools and Manufacture, 48*, 7-8, 832-840. With permission from Elsevier to reprint figures.

[9] Maia, N.M. and Silva, J.M. (1997) *Theoretical and Experimental Modal Analysis.* Research Studies Press Ltd., Somerset, U.K.

[10] NAFEMS (1992) *A Finite Element Dynamics Primer.* Edited NAFEMS, Glasgow, U.K.

[11] Heylen, W., Lammens, S. and Sas, P. (1997) *Modal Analysis Theory and Testing.* KUL Press, Leuven, Belgium.

[12] Haranath, S., Ganesan, N. and Rao, B.V.A. (1987) Dynamic analysis of machine tool structures with applied damping treatment, *International Journal of Machine Tools and Manufacture, 27*, 1, 43-55.

[13] Sharan, A.M., Sankar, S. and Sankar, T.S. (1981) Dynamic behaviour of lathe spindles with elastic support including damping by finite element analysis, *Shock Vibration Bulletin, 51*, 83-96.

[14] Lin, Y.H. (1990) Dynamic modeling and analysis of a high speed precision drilling machine, *Journal of Vibration and Acoustics, ASME, 112*, 3, 355-365.

[15] Avilés, R. and Ajuria, M.B. (1995) *Análisis Dinámico Mediante Elementos Finitos.* Faculty of Engineering of Bilbao.

[16] Erfurt, F, and Tietz, W. (1987) The application of modal analysis in connection with the finite-element method to determine the dynamic behaviour of machine tools, *Proceedings of the 5th International Modal Analysis Conference*, 766-771, London, U.K.

[17] Allemang, R.J. and Brown, D.L. (1982) A correlation coefficient for modal vector analysis, *Proceedings of IMAC, 1*, 110-116. Orlando, U.S.

[18] Ahmadian, H., Mottershead, J.E. and Friswell, M.I. (1996) Joint modelling for finite element model updating, *Proceedings of IMAC, 14*, 591-596, Dearborn, U.S.

[19] Yuan, J.X. and Wu, X.M. (1985) Identification of the joint structural parameters of machine tool by DDS and FEM, *Journal of Engineering for Industry, ASME, 107*, 64-69.

[20] Friswell, M.I. and Mottershead, J.E. (1995) Finite Element Model Updating in Structural Dynamics, *Kluwer Academic Publishers*. Dordrecht, Netherlands.

[21] Dynamic Design Solutions N.V. (DDS), FEMtools Theoretical Manual, Version 3.0, Leuven, Belgium, 2004.

[22] Preumont, A. (2002) *Vibration Control of Active Structures, An Introduction*, Kluwer Academic Publishers, Dordrecht, Netherlands.

[23] Qu, Z-Q (2004) Model Order Reduction Techniques, with Applications in Finite Element Analysis. *Springer*, London, U.K.

[24] Marinescu, I.D., Hitchiner, M., Uhlmann, E., Rowe, W.R. and Inasaki, I. (2007) *Handbook of Machining with Grinding Wheels.* CRC Press, Florida, U.S.

[25] Gawronski, W.K. (2004) *Advanced Structural Dynamics and Active Control of Structures*, Springer, New York, U.S.

[26] Moore, B.C. (1981) Principal component analysis in linear systems: controllability, observability, and model reduction, *IEEE Transactions on Automatic Control, 26*, 1, 17-32.
[27] Hatch, M.R. (2001) *Vibration Simulation using Matlab and Ansys*, Chapman and Hall, CRC Press, Florida, U.S.
[28] Fernando, K.V. and Nicholson, H. (1982) Singular perturbational model reduction of balanced systems, *IEEE Transactions on Automatic Control, 27*, 2, 466-468.
[29] Olgac, N. and Hosek, M. (1998) A new perspective and analysis for regenerative machine tool chatter, *International Journal of Machine Tools and Manufacture, 38*, 7, 783-798.
[30] Chapra, S.C. and Canale, R.P. (2003) *Métodos Numéricos para Ingenieros*, McGraw-Hill Interamericana, México.
[31] Albizuri, J., Fernandes, M.H., Garitaonandia, I., Sabalza, X., Uribe-Etxeberria, R. and Hernández, J.M. (2007) An active system of reduction of vibrations in a centerless grinding machine using piezoelectric actuators, *International Journal of Machine Tools and Manufacture, 47*, 10, 1607-1614. With permission from Elsevier to reprint figures.

In: Machine Tools: Design, Reliability and Safety
Editor: Scott P. Anderson, pp. 81-116
ISBN: 978-1-61209-144-0
© 2011 Nova Science Publishers, Inc.

Chapter 3

LIGHT MACHINE TOOLS FOR PRODUCTIVE MACHINING

J. Zulaika[1]∗, F. J. Campa[2]• and L. N. Lopez Lacalle[2]♦

[1] Fundacion Fatronik, San Sebastian, Spain
[2] Department of Mechanical Engineering,
University of the Basque Country, Bilbao, Spain

ABSTRACT

This chapter presents an integrated 'machine and process' approach that is based on stability lobe diagrams for designing large-volume milling machines, bearing in mind their productivity, reliability and accuracy as well as their eco-efficiency. In fact, eco-efficiency of machining processes is an issue of increasing concern among both machine tool builders and manufacturers. Thus, this chapter introduces a global modeling of the dynamics of milling machines and of milling processes with a goal of supporting engineers in the design of machine tools that result in optimal machining productivity at minimized environmental impacts and costs. This is a result of reducing the machines' material content, and consequently resulting in lowered energy consumption. This approach has been applied to the design of an actual milling machine, upon which machining tests have been conducted, that have shown an increase of 100% in productivity while consuming 15% less energy. This result is due to a weight reduction of over 20% in structural components, thus integrating highly productive machining processes and eco-efficient milling machines in one unique system.

∗ Paseo Mikeletegi 7, E-20009 Donostia- San Sebastian, Spain. e-mail: jzulaika@fatronik.com.
• Escuela Técnica Superior de Ingenieros Industriales, c/Alameda de Urquijo s/n, E-48013 Bilbao, Spain, E-mail: fran.campa@ehu.es.
♦ E-mail: norberto.lzlacalle@ehu.es.

1. INTRODUCTION

The *eco-efficiency* of a machine tool can be defined as the ratio between the extent to which its functional objectives have been met combined with the environmental impact associated with its total lifecycle. In quantitative terms, machine eco-efficiency can be expressed as the ratio between the productivity of the machine and the environmental impact associated with its lifecycle.

With the aim of measuring the *environmental impact* associated with the lifecycle of a machine tool, the authors carried out a *Life Cycle Assessment* (LCA) on a milling machine made by "Nicolas-Correa" Company, considering a life span of 10 years for the machine. When conducting the LCA, the authors followed ISO 14040 rules and selected the "Climate change" impact category within the Ecoindicator 99 methodology [1]. The main conclusion that the authors reached from the LCA was that much of the environmental impact of a machine tool – around 95% - was associated with its use phase. Furthermore, within the use phase, 95% of its impact was produced by the energy that the machine consumed. Figure 1 shows this data both graphically and numerically.

The *productivity* associated with a machine tool or with a machining process is usually measured in terms of *Material Removal Rate* (MRR), which is defined as the volume of material removed per unit of time. The MRR, usually referred to as Z_w, is calculated as the product between the cutting area and the feed velocity of the tool. In the great majority of cases, this productivity is limited by the appearance of self-induced vibrations, vibrations that tend to be commonly referred to as *chatter*. Among the types of self-excited vibrations, attention should be paid to *regenerative chatter* [2], due to both the frequency with which it occurs and its intensity. This is the self-excited vibration *par excellence* in the machining process, and is caused by the regeneration of the metal thickness in systems in which the blade cuts a previously machined surface either totally or partially [3]. This phenomenon will be analyzed in detail in the next paragraphs.

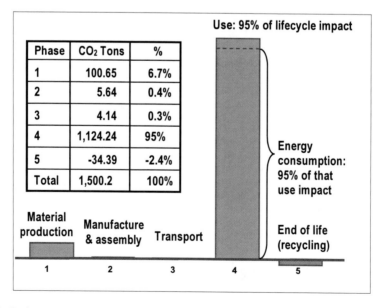

Figure 1. Life Cycle Assessment Study on a milling machine [1].

This chapter will integrate the two concepts of eco-efficiency and productivity, that traditionally have been viewed as being mutually exclusive goals. Currently, the standard practice for machine tool builders is to produce high-performance machines by means of high dynamic feed-drives [4], which forces design engineers to utilize heavy and stiff mechanical structures capable of bearing the higher inertial forces. The increased weight of structural elements in turn, creates a requirement for more powerful actuators, leading to higher inertial forces. This is a vicious circle in which the increase of productivity in machining processes has been achieved at the expense of high material and energy consuming machines, increasing both the environmental impacts and costs associated with machining processes.

Towards a goal of building machine tools that are both highly productive and eco-efficient, this chapter presents an integrated 'machine and process' approach that implements separating the productivity of machining processes from the material intensity of machine tools. The details of this approach are based on achieving a specific value of productivity in machining processes by constructing machines using the minimum amount of material in their structural elements, leading to a reduction in both the material resources consumed in the production phase and in the energy resources consumed by the machines during their use phase.

The conceptual cornerstone for this machine *dematerialization* process lies in defining the exact dynamic and mechanical performances that are required for the machine to assure specific productivity goals, and achieving them with the least possible weight of mobile structural elements.

At this point, it is worthwhile to reflect on the functionality of the structural elements of a machine tool. On the whole, the structure of a machine tool has two main functions [5]:

1. To hold the components and peripherals involved in the machine and to assure their kinematic requirements during machine motions and processes.
2. To withstand the forces that are produced by the machining process and by the machine motions in a robust manner for assuring the required motion accuracy.

Concerning these two functions, it is feasible to fulfil the former requirement - i.e. to hold machine components and provide the required kinematic performances - with structural components of much lower weight. An example of this is the case of industrial robots, which are much lighter than machine tools. In quantitative terms, according to a study of the authors, it is possible to assure the kinematic and dynamic performances of machine components with approximately 25% of the weight of the mechanical structure of an average milling tool [6]. This means that the functionality of the remaining 75% of the mechanical structure is only to assure robust performance of the machine during machine positioning and production processes. Therefore, for a specific machining process, if the specifics of the minimum requirements for stiffness and robustness are identified and quantified, and they are achieved using a minimum of structural weight, the result is a wide range of possibilities for reducing the weight of a machine in a significant manner without affecting its productivity.

Following this conceptual approach, the next subchapters will analyze on one hand, the mechanical design requirements associated with specific values of process productivity and accuracy, and on the other hand, will analyze synergetic and complementary approaches for assuring process productivity while using a minimum of material in the machine structure.

Table 1. Average forces and acceptable deformations for different milling operations

Process	Average max. force on TCP	Acceptable deformation
Roughing	Conventional tools: 1,500 N 100-125 mm. diam. tools: 3,000 N	Average: 100 µm
Semi-finishing	Conventional tools: 1,000 N	Average: 50 µm
Finishing	Conventional tools: 200 N	Average: 10 µm

2. DESIGN REQUIREMENTS FOR ACHIEVING SPECIFIC PRODUCTIVITY LEVELS

With the aim of defining the design requirements that lead to productive and stable machining processes, the mechanical requirements that are associated with both static and dynamic stiffness of the machine tool structures will be analyzed.

2.1. Requirements of Static Stiffness for Productive and Accurate Machine Tools

Table 1 shows the average process forces associated with milling operations when cutting AISI 1045 Steel [7]. This data has been measured experimentally by the authors on various machines in different workshops, and can be used as a basis for defining threshold values for static stiffness of a machine tool at the *tool center point* (TCP) that will assure an accurate and reliable performance of the machine. These results have been combined with accuracy requirements that machine end-users have provided for the different milling operations, which has allowed the authors to develop operational specifications associated with different machining operations.

Combining the actual forces that have been gathered in the experimental measurements with the precision requirements that the end users have provided to the authors, and taking into account that the main component of machining forces is in the tangential direction of the tool, a static stiffness of 1500N/100µm (i.e. 20 N/µm) at the tool center point in the tangential direction can be considered as a reference threshold value for general-purpose milling machines. This means that if a machine has a higher static stiffness, from the static point of view there is margin for reducing the weight of the machine without affecting the accuracy and reliability of machining processes.

2.2. Requirements of Dynamic Stiffness for Productive and Reliable Machine Tools

In addition to the requirements associated with the static stiffness of a machine tool, it is worthwhile to highlight that in the great majority of machining processes, the productivity of machining processes is not limited by static deformations but by the presence of vibrations, in particular of self-induced vibrations. Among these self-induced vibrations, special attention should be paid to *regenerative chatter,* owing both to its incidence and to its intensity. This regenerative chatter is caused by the regeneration of the metal thickness in systems in which the blade cuts a previously machined surface either totally or partially, and its appearance is linked to the following parameters [8, 9]:

i) The dynamic stiffness of the machine-process system;
ii) The geometry of the machining tool;
iii) The radial immersion of the machining tool into the material to be removed
iv) The material to be removed.

From this perspective, if the part to be machined is of a hard material such as steel, and its dimensions are below the dimensions of the machine, the dynamic performance of the system will be dominated by the natural frequencies of the machine structure, as shown in Figure 2 [10].

As can be seen in Figure 2, the natural frequencies of the machine-process system at low frequencies (below 200 Hz.) are related to structural natural frequencies of the machines, whereas the natural frequencies above 200 Hz. are related to natural frequencies of machine components such as spindle, tool, tool-holder etc. The manufacturers distinguish between two general types of chatter, the so-called 'structural chatter' or 'machine chatter', which is a low frequency chatter, and the so-called 'tool chatter', which appears at higher frequencies. In the former case, recognizable as a low-pitched sound, the chatter is associated with the structural modes of the machine, whereas in the latter case, marked by a high-pitched sound, the chatter is associated with the modes of specific machine components such as spindle, tool etc.

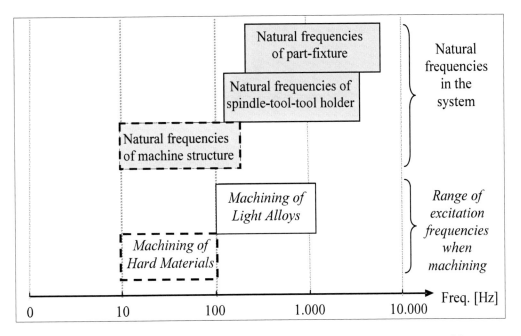

Figure 2. Ranges for natural frequencies of the system *(upper part, grey-shaded boxes)* and for excitation frequencies when machining different materials *(blank boxes)* [10].

The type of chatter that will be excited by the machining forces will depend on the excitation frequencies associated with those machining processes. According to Figure 2, when machining hard materials such as steel, the associated excitation frequencies will be in the range of the natural frequencies of the structural elements of the machine. Therefore, the stability of this type of machining processes will be affected mainly by the modal parameters associated with the *machine structure*.

Within this view, in the next subchapter, a stability model adapted to the machining of parts of hard materials that are smaller than the machine dimensions will be developed, which will serve as a basis for translating process stability requirements into design requirements in terms of dynamic stiffness at the TCP.

2.2.1 Stability Model for the Machining Process

As a starting point, a stability model has been developed for milling processes that is based on a mechanistic modeling of the milling forces [8], a mono-frequency approach of the stability solution [11], the introduction of the dynamics of the machine tool in modal coordinates as in [12], and the influence of the feed direction as in [13]. In addition, the modal vector at the tool tip is referenced to the machine tool axes X_{MT} Y_{MT} Z_{MT} and to the local axes of the tool X_t Y_t Z_t, coinciding Z_{MT} and Z_t axes and defining X_{MT} and Y_{MT} the working plane of the machine and $+X_T$ the feed direction of the tool, as shown in Figure 3. $\bar{i}_m$ is the versor that defines the direction of a mode and $\{\phi_i\}$ the modal vector per mass unity, the projection of the versor over the $X_{MT}Y_{MT}$ plane is $\bar{i}_m^{xy}$, whose angle with respect to the feed direction is β_{xy}, whereas the angle between the modal vector and the $X_{MT}Y_{MT}$ plane is β_z. Hence, the direction of a given machine mode is related to the feed direction $(+X_T)$ through the two angles β_{xy} and β_z.

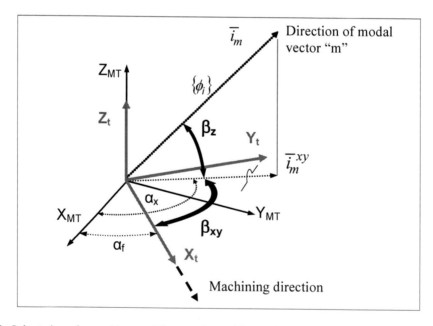

Figure 3. Orientation of a machine modal vector in machine tool axes *(MT indexes)* and tool axes *(t indexes)*.

For a jth cutting edge of a milling tool, the mechanistic cutting forces model proposes a linear relation between the dynamic chip thickness h_d and the width of cut b in the cutting area (see Figure 4 below right) with the tangential, radial and axial forces F_t, F_r and F_a (see Figure 4 below) that are acting over the cutting edge, being the linear relation based on tangential, radial and axial *cutting coefficients* K_{tc}, K_{rc} and K_{ac} respectively:

$$F_{tra,j}(\phi_j) = \begin{Bmatrix} F_{t,j} \\ F_{r,j} \\ F_{a,j} \end{Bmatrix} = \begin{Bmatrix} K_{tc} \\ K_{rc} \\ K_{ac} \end{Bmatrix} \cdot b \cdot h_d(\phi_j) = \frac{K_t \cdot a_p}{sin\kappa} \begin{Bmatrix} 1 \\ K_r \\ K_a \end{Bmatrix} h_d(\phi_j), \qquad (1)$$

where ϕ_j is the angular position of that jth cutting edge related to the $+Y_t$ axis (see Figure 4 below left), and the width of cut b can be related to the axial depth of cut a_p by means of the cutting edge lead angle of the insert κ, $b=a_p/sin\kappa$ (see Figure 4 below right). In addition, for the sake of clarity, the cutting coefficients K_{tc}, K_{rc} and K_{ac} can be normalized with respect to the tangential component $K_t=K_{tc}$, as shown in equation (1), being $K_r=K_{rc}/K_{tc}$ and $K_a=K_{ac}/K_{tc}$.

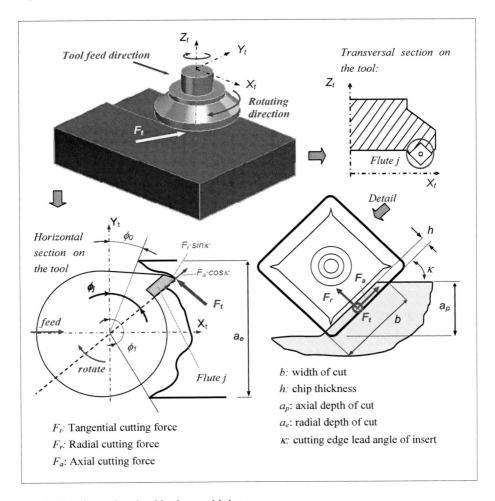

Figure 4. Cutting forces involved in the machining process.

By means of this mechanistic model for the machining forces and a dynamic model of the machine, the dynamic machining forces that are acting on "j" flute can be projected over the mth mode direction $\overline{i_m}$, calling a_F^i to those projections. Bearing in mind the modal vector modulus ϕ_i for the ith mode and considering n modes of the machine, the process forces can be expressed in modal coordinates g_m:

$$\begin{Bmatrix} g_{1,j} \\ g_{2,j} \\ \vdots \\ g_{n,j} \end{Bmatrix} = [\{\phi_1\} \ \{\phi_2\} \ \cdot \cdot \ \{\phi_n\}]^T \cdot \begin{bmatrix} a_{F_t}^1 & a_{F_r}^1 & a_{F_a}^1 \\ a_{F_t}^2 & a_{F_r}^2 & a_{F_r}^2 \\ \vdots & \vdots & \vdots \\ a_{F_t}^n & a_{F_r}^n & a_{F_r}^n \end{bmatrix} \cdot F_{tra,j} \cdot \qquad (2)$$

On the other hand, each of these "n" modes has an associated modal displacement in the mode direction that will be called δ. In this respect, naming as Δm_i the dynamic deformation in global coordinates that is associated with a mode "i" of the machine at the tool centre point, that displacement will be expressed as a modal displacement $\Delta \delta_i$ by means of the following expression, where $\{\phi_i\}$ is the modal vector per mass unity for that mode "i" at the tool center point:

$$\Delta m_i = \phi_i \Delta \delta_i . \qquad (3)$$

Generally, the change from global coordinates y_i into modal coordinates δ_i is conducted by means of the following transformation in the Laplace domain:

$$\{\delta(s)\} = [\Phi]\{y(s)\}. \qquad (4)$$

where $[\Phi]$ represents the matrix of modes, whose columns are the "n" eigen-modes of the considered mechanical system:

$$[\Phi] = [\{\phi\}_1, \{\phi\}_2, \cdots \{\phi\}_j, \cdots \{\phi\}_n]. \qquad (5)$$

Thus, in a machine where "n" modes have been considered, the effect of each of the modal displacements $\Delta \delta_i$ h_d on the dynamic chip thickness will become:

$$h_d(\phi_j) = \{\phi_1 a_{h_1} \quad \phi_2 a_{h_2} \quad \cdots \quad \phi_n a_{h_n}\} \begin{Bmatrix} \Delta \delta_1 \\ \Delta \delta_2 \\ \vdots \\ \Delta \delta_n \end{Bmatrix}. \qquad (6)$$

where the terms a_{h_i} represent the projection of the modal displacements in the direction of the chip thickness h_d, as shown in Figure 5, and takes the following expression:

$$a_{h_i} = (sin\phi_j cos\beta_{xy} + cos\phi_j sin\beta_{xy})cos\beta_z \cdot sin\kappa - cos\kappa \cdot sin\beta_z. \quad (7)$$

Now that both forces and displacements have been expressed in modal coordinates, integrating the former equations, the following relation is achieved with respect to modal forces and displacements in a machining process [14]:

$$g = K_t \cdot a_p \cdot A(\phi) \, \Delta\delta/2, \quad (8)$$

where K_t represents the tangential cutting coefficient, a_p the axial depth of cut, $\Delta\delta$ the dynamic displacement of the tool in modal coordinates and $A(\phi)$ a matrix of dimensionless factors.

Matrix $A(\phi)$ consists of dimensionless factors called *directional factors*, where for the row "p" and the column "q", the element a_{pq} of the matrix takes the following expression:

$$a_{pq} = \frac{2 \cdot \phi_q \cdot a_{h_q} \cdot \phi_p}{sin\kappa}\left[a_{F_t}^p + K_r \cdot a_{F_r}^p + K_a \cdot a_{F_a}^p\right]. \quad (9)$$

According to equation (2), a term a_F^i represents the projection of a force F on the direction of a mode "i", whereas according to equation (7), a_{h_i} represents the projection of a modal displacement that is along the direction of a mode "i" on the direction of the dynamic chip thickness h. Thus, a a_{pq} element of that matrix indicates to which extent a dynamic displacement along the direction of a mode "p" affects the chip thickness h and by extension, the generation of process forces that excite the mode "q". This coefficient can be either positive or negative; a positive coefficient means that the machining forces make the tool separate from the workpiece, whereas a negative coefficient means that the machining forces make the tool penetrate the workpiece.

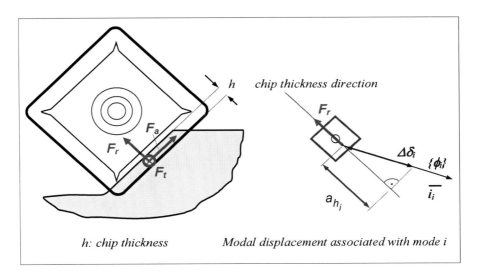

Figure 5. Effect of modal displacements on the dynamic chip thickness h_d.

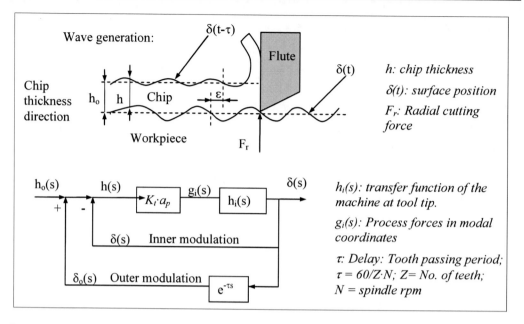

Figure 6. Block diagram of chatter dynamics [15].

When the cutting forces excite any of the structural modes of the machine that is holding the tool, a wavy surface finish is left due to structural vibrations. That wavy surface left by the previous tooth is removed during the succeeding tooth period, which also leaves a wavy surface due to the dynamic cutting forces [15]. Thus a self-excitation mechanism is generated during the material removal process, in which the output process variable –the position of the tool- affects the input process variable –the chip thickness-, generating a positive feedback system.

On the other hand, depending on the phase shift between two successive waves, the chip thickness may grow exponentially while oscillating at a chatter frequency ω_c that will be close to the dominant structural mode of the machine, i.e. the natural frequency of the open loop system in Figure 6.

Based on the block diagram that is shown in Figure 6, a mono-frequency solution is applied to it to solve its characteristic equation. Thus, if chatter appears at a frequency ω_c, the dynamic displacement $\Delta\delta_i$ of an ith mode in modal coordinates and for a tooth passing period τ can be obtained from the dynamic cutting forces and from its modal parameters. Namely, the involved parameters are its natural frequency ω_i, its modal vector ϕ_i, its damping coefficient ξ_i and its effective stiffness $k_{efi}=\omega_i^2/\phi_i^2$, and the displacement takes the following expression [14]:

$$\Delta\delta_i = (1 - e^{-i\omega_c\tau}) \cdot h_i(\omega) \cdot g_i, \qquad (10)$$

in which h_i is the transfer function of the machine at the TCP in modal coordinates. If the variable $r_i=\omega_c/\omega_i$ is defined, the transfer function h_i in modal coordinates takes the following form:

$$h_i(\omega) = \frac{1/\omega_i^2}{(1-r_i^2)+i\cdot(2\xi_i r_i)^2} \cdot \qquad r_i = \omega_c/\omega_I \qquad (11)$$

Thus, for the case of a machine with "n" structural eigen-modes, the total dynamic displacements of the "n" modes will take the following form:

$$\Delta\delta = (1-e^{-i\omega_c\tau})\cdot H(\omega)\cdot g, \qquad (12)$$

in which H(ω) is a diagonal matrix nxn in which the elements of the diagonal are the transfer functions h_i in modal coordinates:

$$H(\omega) = \begin{bmatrix} h_1(\omega) & 0 & \cdots & 0 \\ 0 & h_2(\omega) & \cdots & 0 \\ \vdots & \vdots & \ddots & \vdots \\ 0 & 0 & \cdots & h_n(\omega) \end{bmatrix}. \qquad (13)$$

If the previous equations are combined and the mean Fourier term of the directional coefficient matrix is adopted, naming α the resulting matrix, the *eigen-equation* (14) is reached, in which Z is the number of flutes on the tool:

$$g = \frac{Z}{4\pi} K_t \cdot a_p \cdot (1-e^{-i\omega_c\tau})\cdot \alpha\cdot H(\omega)\cdot g. \qquad (14)$$

2.2.2. Stability Problem Solution

The solution of the *nth* order eigen-value problem in modal coordinates shown in (14) provides the relation between the chatter frequency, the critical depth of cut and the spindle speed, and allows the assembling of the stability lobes and chatter frequency diagrams, as in [11].

As the equations are expressed in modal coordinates, they are decoupled among them except for the case of the α matrix of directional factors, which is not symmetrical. If the asymmetry of the α matrix is neglected, each mode can be analyzed in an individual manner.

Thus, for equation (14), for the specific case of considering *one concrete structural mode "q" of the machine* (actually, in the real machining cases, for one specific machine position and for one specific machining direction, there will always be *one* predominant eigen-mode of the machine that will cause process instabilities), and finding the depth of cut a_p in that expression, the resulting value is the following:

$$a_p = \frac{2\pi}{Z\cdot K_t}\cdot\frac{1}{\alpha}\cdot\frac{1}{real(h(\omega))}, \qquad (15)$$

where α is the *directional coefficient* associated with the considered eigen-mode "q". Basing on equation (9), the average Fourier term for the directional factor associated with that mode "q" takes the following form:

$$a_q = \frac{2}{\sin\kappa} \phi_q^2 \cdot \int_{\phi_0}^{\phi_1} a_{h_q}(a_{F_t}^q + K_r \cdot a_{F_r}^q + K_a \cdot a_{F_a}^q) d\phi . \tag{16}$$

Coming back to equation (15), for the specific case of a mechanical system of one degree of freedom, the real part of a transfer function with a mechanical stiffness k and a modal damping ξ takes the form that is shown in Figure 7.

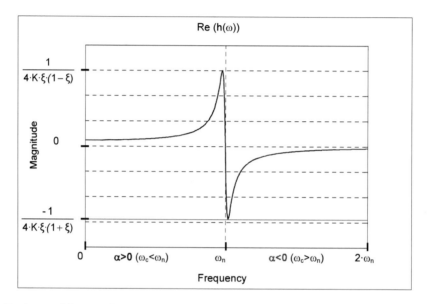

Figure 7. Real part of the transfer function of a system of one degree of freedom.

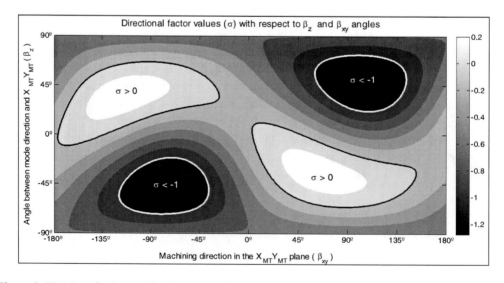

Figure 8. 2D Map of values of the directional factor σ with respect to β_{xy} and β_z angles.

As can be seen in Figure 7, the maximum and the minimum values of the real part of the transfer function takes the following form:

$$\max/\min(Re(h(\omega))) = \frac{1}{4\cdot k\cdot \xi\cdot(1\mp\xi)}. \qquad (17)$$

On the other hand, a new term σ can be defined as the *real directional factor*, for removing the modal vector ϕ_q in the expression of the *directional factor* α_q:

$$\sigma = \frac{\sin\kappa}{2\cdot\phi_q^2}\cdot\alpha_q. \qquad (18)$$

Thus combining the equations (15) to (18) and finding the expression of the critical depth of cut a_{pcrit}, the following expression is obtained (19):

$$\sigma < 0 \rightarrow a_{p_{crit}} = \frac{4\pi\cdot\sin\kappa}{Z\cdot K_t}\cdot\frac{1}{\sigma}\cdot\frac{k_{ef}\xi(1+\xi)}{1} \cong \frac{4\pi\cdot\sin\kappa}{Z\cdot K_t}\cdot\frac{1}{\sigma}\cdot k_{ef}\cdot\xi.$$

$$\sigma > 0 \rightarrow a_{p_{crit}} = \frac{4\pi\cdot\sin\kappa}{Z\cdot K_t}\cdot\frac{1}{\sigma}\cdot\frac{k_{ef}\xi(1-\xi)}{1} \cong \frac{4\pi\cdot\sin\kappa}{Z\cdot K_t}\cdot\frac{1}{\sigma}\cdot k_{ef}\cdot\xi. \qquad (19)$$

k_{ef} is the effective stiffness of the considered mode ($k_{ef}=\omega_n^2/\phi^2$, being ϕ the modal vector per mass unit), that in a system of one degree of freedom its value coincides with the value of the static stiffness k, $k_{ef} = k$. Furthermore, κ is the lead angle of the cutting edge and σ is the *real directional factor*, a dimensionless factor that depends on the orientation between the mode and the feed direction as well as on the normalized radial and axial cutting coefficients [13].

Figure 8 shows a 2D map of the values that this coefficient can reach with respect to β_{xy} and β_z angles. If $|\sigma|<1$ this coefficient increases the value of the modal effective stiffness k_{ef}, and in turn, if $|\sigma|>1$ it decreases that value.

Finally, it must be noted that in equation (19), an approximation of $(1+\xi) \approx 1$ has been applied, since the damping coefficient ξ that is associated with structural models of machine tools is negligible compared with a value of unity. Indeed, the authors have measured an average value of $\xi=0.02$ for structural modes of several milling machines of different architectures.

2.2.3. Requirements of Dynamic Stiffness for a Specific Productivity Level

As the productivity of a machining process is directly proportional to the depth of cut a_p, according to (19), to obtain a specific level of productivity in a machine tool in terms of MRR, the specific structural mode of the machine that limits the stability of the machining process requires a threshold value for the product of the *effective stiffness* k_{ef} and the *damping coefficient* ξ associated with that limiting mode. It is also notable that the real directional factor σ may increase or decrease the actual value of the modal effective stiffness.

Thus, coming back to (19), considering the machine tool as a system of one degree of freedom with associated effective stiffness k_{ef} and damping factor ξ, and defining a factor V as V=((Z·K$_t$)/(4·πsinκ)), the *dynamic requirement* associated with the *productivity of the machining process, per millimeter of depth of cut* a_p, turns out to be:

$$\frac{1}{\sigma} \cdot k_{ef} \cdot \xi \geq V . \qquad (V = Z \cdot K_t / 4 \cdot \pi \sin \kappa) \qquad (20)$$

2.3. Achieving Productivity Targets with the Least Possible Weight

This integral machine and process approach aims at achieving the critical depth of cut shown in (19) with a minimum of material content in the machine. The proposed approach is that if a specific machine mode does not limit the productivity of the machine, its modal stiffness can be reduced without affecting the productivity and stability of the machining process. And if one specific mode limits the productivity of the machine in one specific milling direction at a specific speed, according to (19) and (20), the productivity of the machine can be maintained or even improved while reducing the mass of the machine by means of the following two complementary approaches:

1) By modifying the real directional factor σ for reducing its absolute value.
2) By reducing in a deliberated manner the modal stiffness associated to that mode – with a subsequent reduction in the mass of involved machine components - and by increasing, in parallel, the modal damping in the same proportion by means of active damping devices, thus maintaining constant the product $k_{ef} \cdot \xi$.

Next, these two complementary approaches for assuring a specific value of productivity with a minimum of machine weight will be analyzed.

2.3.1. Modification of the Directional Factor σ for Reducing the Need of Machine Weight for Specific Productivity Levels

The *real directional factor* σ is a dimensionless factor that depends on the relative orientation between the feed velocity of the tool and the modal vector and on the cutting coefficients normalized with respect to the tangential coefficient. For one mode "q" of the machine, according to equations (16) and (18), it takes the following form:

$$\sigma_q = \int_{\phi_0}^{\phi_1} a_{h_q} (\cdot a_{F_t}^q + K_r \cdot a_{F_r}^q + K_a \cdot a_{F_a}^q) d\phi \qquad (21)$$

where a_F^q refers to the projection of a force F on the direction of a mode "q", and a_{h_q} refers to the projection of the modal displacement associated with mode "q" on the direction of the chip thickness h_d. Low absolute values of this factor improve the stability of the process, since the critical depth of the cut will tend to increase, as can be deduced from (19).

As it has been shown in Figure 8, this directional factor σ varies in a notable manner with respect to angles $β_z$ and $β_{xy}$ (angles between the mode and the working plane and between the projection of the mode on the working plane and the tool feed direction, respectively). As a reference, Figure 9 shows how this directional factor varies on the working plane for different $β_z$ angles,

Figure 9 shows that directional factors oscillate in a notable manner with respect to the values of $β_{xy}$ angle, and the fluctuation characteristics depend on the value of $β_z$ angle: for $β_z$= 0°, oscillations are low, for $β_z$= 45° oscillations are more pronounced, and for $β_z$= 90° those oscillations disappear, with constant factors.

Figure 10a shows, in turn, the worst case of σ from the process stability point of view, that is the maximum absolute value, for different $β_z$ angles (having swept $β_{xy}$ from 0° to 360° for each of the $β_z$ angles). Finally, Figure 10b shows the average value of absolute values of σ for different $β_z$ angles (having swept $β_{xy}$ from 0° to 360° by steps of 1°).

Figure 10a shows that in the range of $β_z$ angles from 22° to 72°, there are $β_{xy}$ angles, i.e. machining directions in the $X_{MT}Y_{MT}$ working plane, where the values of the directional factors are above unit in absolute value, |σ|>1, which leads to reduced values of effective stiffness. The same information can be extracted from the 2D map that has been shown in Figure 8 for a specific $β_z$ angle. In calculating average values of directional factors when sweeping the range of machining directions from 0° to 360° with a resolution of one degree, it turns out that for increasing values of $β_z$ angles, the average value of the directional factor, in absolute value, also increases.

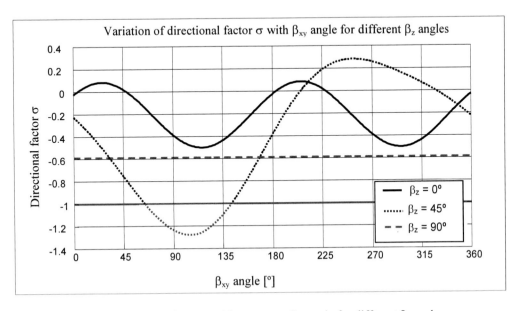

Figure 9. Variation of directional factor σ with respect to $β_{xy}$ angle for different $β_z$ angles.

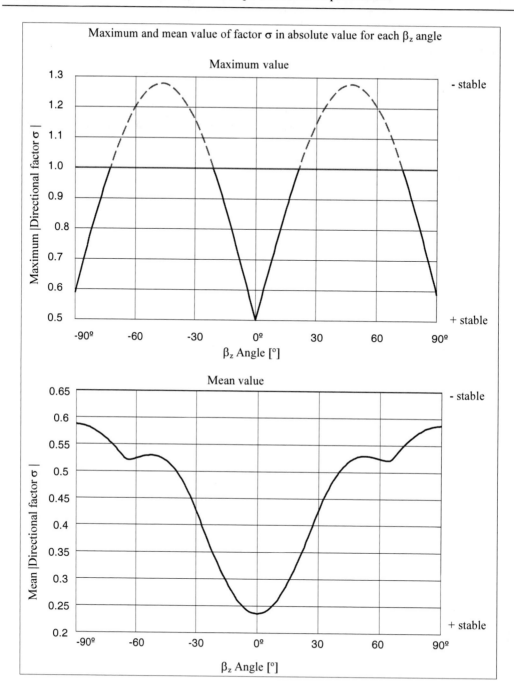

Figure 10. a) Above: Maximum absolute value of directional factor σ for different β_z angles. b) Below: Average of absolute values of directional factor σ for different β_z angles.

Therefore, it can be concluded that for the cases where there are predominant machining directions, e.g. roughing of large parts in the longitudinal direction, the machine can be designed so that their structural modes and the predominant machining directions do not lead to combinations where directional factors are above unit. For the cases that there are not predominant machining directions, the general rule is that for the foreseen dominant modes, the lower the β_z angle, the more stable the machining process, as shown in Figure 10b. That means that for the case that the modal vector is contained in the $X_{MH}Y_{MH}$ working plane, i.e. $\beta_z=0°$, the machining processes will be more stable, or in other terms, there will be margin during the design stage for reducing the machine stiffness -and consequently, the material intensity of the machine- for an aimed value of machining productivity.

This approach is not easy to be applied in practice, since mode directions are highly dependant of machine architecture, so that changing the direction of modes is not a trivial issue. One interesting alternative to changing the direction of modes lies in changing the machine table orientation, as shown in Figure 11, because this way, the angle between the considered machine mode and the working plane, i.e. the β_z angle, is changed without changing the machine architecture.

As a reference, Figure 11 shows a milling machine with an architecture based on a mobile column and an embedded horizontal ram. For that machine architecture, it is quite common that the predominant mode is the bending mode in the vertical direction of the horizontal ram, as shown in Figure 11 above, so that the modal vector associated with that mode will be approximately a vertical vector. On the other hand, it must be noted that the X_{MT} and Y_{MT} axes are contained within the working plane, so that if the working plane is re-oriented, the directions of the X_{MT} and Y_{MT} axes are also modified, so that the vector components of the bending mode have to be adapted.

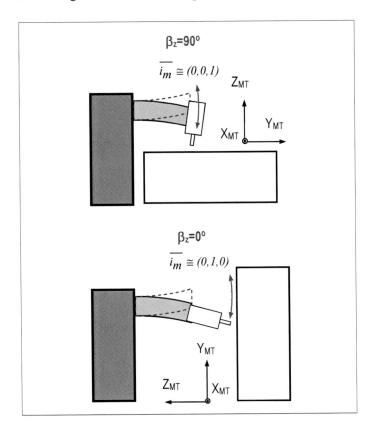

Figure 11. Change in the working plane orientation for modifying the directional factor σ and thus the process productivity. *Above*: Initial machine architecture; *Below*: re-designed machine architecture.

In this case, if instead of changing the direction of the predominant mode, the orientation of the $X_{MH}Y_{MH}$ working plane is changed, the angle between the bending mode and the working table will pass from the initial $\beta_z \approx 90°$ to $\beta_z \approx 0°$, i.e. the modal vector associated with the bending mode passes from being normal to the working plane ($\beta_z \approx 90°$) to being contained within the working plane ($\beta_z \approx 0°$).

This means that it is possible to increase the stability of a machining process and the productivity of the machine without modifying the mechanical robustness or the material intensity of the machine structure. Therefore, for a given process productivity goal, there is a possibility of maximizing the directional factor σ by adapting the orientation between the mode and the working plane, thus gaining a margin for reducing the dynamic stiffness associated with the predominant modes and consequently, the mass associated with the machine without affecting its productivity.

2.3.2. Modification of the Product Between Dynamic Stiffness and Damping Coeffiient to Reduce the Need of Machine Weight for Certain Productivity

The second approach to improve the productivity of a machine while reducing the mass of the machine is based on the fact that it is feasible to maintain the critical depth of cut of a machining process while decreasing the stiffness and mass associated with the limiting mode of the machine, provided that in parallel, its modal damping is increased by a set percentage, so that the product $k \cdot \xi$ of stiffness and damping of the limited mode is maintained, as shown in (20).

Within the approach of increasing the modal damping of a limiting mode as a means for gaining margin for reducing the modal stiffness, there are systems that provide additional damping at a range of frequencies or at a given direction, such as viscoelastic materials, viscous fluids and above all, *Active Damping Devices* (ADDs). A typical ADD consists of a vibration sensor, an inertial actuator and a controller. ADDs are based on the principle that an acceleration of a suspended mass results in a reaction force towards the supporting structure. In order to tune the acceleration, an embedded sensor monitors the supporting structure vibration; the sensors readings are sent to an external feedback controller that drives the internal electromagnetic actuator of the ADD. As a result, these devices can damp the vibration modes that they observe in an open-loop transfer function [16].

Figure 12. ADD placed on the ram of a milling machine.

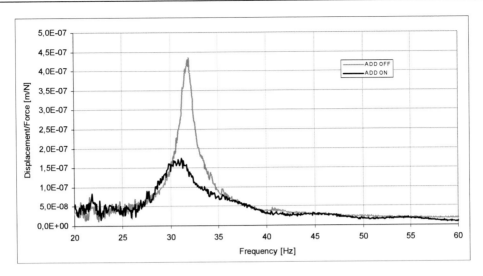

Figure 13. FRF in X_{MT} direction of a horizontal milling machine with and without an ADD.

ADDs add damping to the machine independently of its dynamic properties. The ideal place to locate the ADDs is as close as possible to the tool center point. Thus, in machines that have a movable ram, a good place to allocate ADDs is at the end of the ram close to the headstock, as illustrated in Figure 12.

One interesting aspect related to the use of these ADDs is the estimation of the additional damping that they can provide, so that during the design stage a parallel reduction of the modal stiffness can be implemented for maintaining the product between the stiffness and the damping with a lower material content.

With the aim of quantifying the increase of damping associated with these ADDs, a couple of ADDs of Micromega Company have been placed on a horizontal ram of a milling machine of Fatronik, so that the horizontal bending mode of the machine can be damped, as shown in Figure 13.

Once the ADDs have been placed on the ram of the machine, experimental Frequency Response Functions -FRFs have been measured at the machine TCP, both for the ADD activated and deactivated cases. Figure 13 shows two FRFs in the X_{MT} direction, the first FRF with the ADD activated and the second FRF with the ADD deactivated.

Figure 13 is of great interest, as it shows damping of 60% of the dynamic flexibility associated with the bending mode of the machine, from the initial dynamic flexibility of 4.3e-7 m/N at 32.5 Hz to the final value of 1.7e-7 m/N at 32 Hz. This damping of the mode is equivalent to an *increase of 253% of the relative damping coefficient ξ* associated with that mode in relation to the initial coefficient, considering the system as of one degree of freedom, where the dynamic displacement D_{dyn} is related to the static displacement D_{st} by the following expression:

$$D_{dyn} \approx D_{st} \cdot (1/2\xi). \tag{22}$$

This means that the critical depth of cut for machining processes in which the limiting mode with regard to the appearance of self-induced vibrations is that the damped mode will be around 253% higher with that ADD than the achievable depth of cut without that ADD.

Taking into account that damping coefficient ξ has increased by 253%, it is possible to maintain the product between stiffness and damping coefficient associated with that mode *by decreasing the effective stiffness by 60%*. Thus, a remarkable reduction can be achieved in the material intensity of the machine without affecting its productivity.

3. APPLICATION OF THE APPROACH ON AN ACTUAL INDUSTRIAL CASE

This methodology aimed at integrating productivity and eco-efficiency in milling machines has been applied on an actual industrial case, in concrete on the re-conception of an actual milling machine of a machine tool builder. The selected milling machine has a fixed-table, with a moving column in the longitudinal direction X_{MH}, built-in vertical slide inside the column in the vertical direction Z_{MH} and a built-in horizontal ram inside the slide for the transversal direction Y_{MH}. This machine architecture is commonly known as *box-in-box* type architecture, and is an architecture that is geared towards the general mechanics sector.

The decision for applying the methodology to this family of milling machines has been based on its high ratio between work volume and machine volume, its versatility and, above all, its serial architecture concerning its drives, in the sense that any reduction in weight in an internal axis means a reduction in the weight involved in the external axes. Figure 14 shows an scheme of this machine architecture.

Equipped with a universal milling head, the selected machine reaches feed speeds of 30 m/min and accelerations of 1.2 m/sec^2. Its current column has an approximate height of 3,500 mm and weighs 6.500 kg. The vertical slide contains a section of approximately 1,000 x 1,000 mm^2 and weighs 1,400 kg. The horizontal ram, for its part, contains a section of approximately 500 x 450 mm^2 and weighs 1,450 kg. All these structural elements are made using welded S275 JR steel.

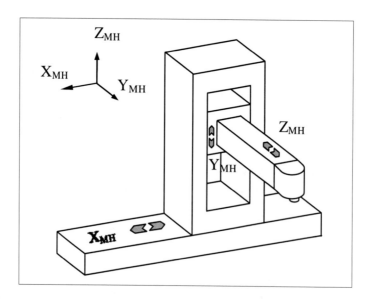

Figure 14. *Box-in-box* architecture of the milling machine to be re-designed.

With regard to the transmission of the milling machine, the axis drive Y in the feed direction of the horizontal ram comprises a spindle of 50 mm in diameter, using a pulley transmission and an asynchronous servo-motor of 10 kW rated output. Axis Z, for its part, comprises a spindle of 63 mm in diameter, with pulley transmission and an asynchronous servo-motor of 10 kW rated output.

3.1. Analytical and Experimental Modal Study of the current Milling Machine

As a starting point for the re-conception of this actual milling machine, an FEM model has been developed, which has enabled the relevant machine modes to be obtained:

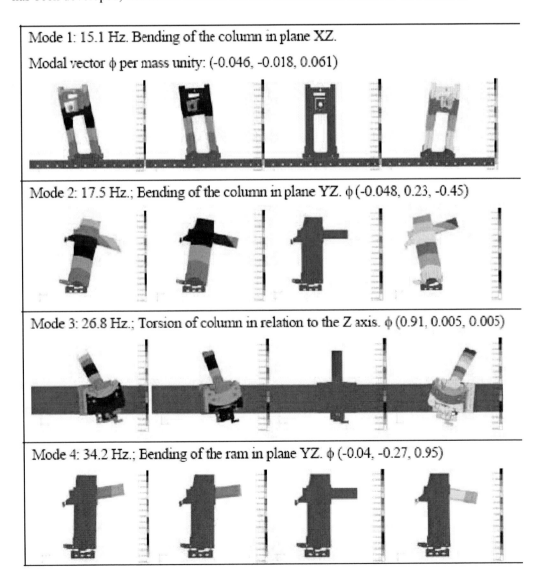

Figure 15. Analytical modal study of the original milling machine to be re-designed.

Table 2. Actual values for damping coefficients ξ

Mode #	Actual value for eigen-frequency	Actual value for damping coefficient ξ
1	13.5 Hz.	1.66 %
2	16.2 Hz.	2.32 %
3	28.5 Hz.	2.66 %
4	36.6 Hz.	3.11 %

This analytical study has been complemented by an experimental modal analysis, which has enabled the real values of the machine's frequencies to be measured and, above all, the real values of the modal damping coefficients.

Next, additional analytical calculations have been conducted on the FEM model of that actual milling machine, with the goal of having as much information as possible about the limiting modes of the machine as well as about the directional factors, within an aim of removing as much weight as possible from the structural components of the machine without affecting its final productivity in terms of MRR.

Thus, Figure 16 shows the values of modal stiffness, effective stiffness and the quotient between the effective stiffness and the maximum absolute value of the real directional factor σ associated with that mode in its working plane.

As it can be seen in Figure 16, the directional factor σ can increase the flexibility associated with a certain machine mode (such as for the case of mode 2), whereas in other cases, it reduces the flexibility associated with the mode (such as for the cases of modes 1, 3 and 4).

On the other hand, based on the calculations that have been shown in Figure 16, the dynamic stiffness associated to these four modes have been calculated in the three machine axes X_{MT}, Y_{MT} and Z_{MT}, showing the results in Figure 17.

As it can be seen from Figures 16 and 17, there is not a direct relation among the modes with highest effective stiffness divided by the directional factor σ on the one hand and the modes with highest dynamic flexibility on the other hand. This aspect is of great interest for the re-design process for milling machines, since it allows joining weight reduction strategies (linked to the dynamic stiffness of a machine) and productivity increase strategies (linked to the effective stiffness divided by the directional coefficient σ).

Next, the stability lobe diagrams have been calculated for these FEM model of the machine and for a typical milling operation, with the aim of identifying *which is the mode or modes that currently limits the productivity of the machine* in given machine positions and machining directions.

Thus, for calculating the stability lobes, the following machining process has been considered: an AISI 1045 steel rough milling operation with a 125 mm plate, 9 flutes at a 45° angle and with 80% radial immersion in concordance, and a working range of 400 to 600 rpm for the tool.

Figure 18 shows a polar diagram that shows the critical depths of cut that have been obtained in this machine for the machining process mentioned above and with the horizontal ram in an outer and lower position. In that polar plot, 0° coincides with the $+X_{MT}$ direction of the machine, and the 360° cover the different machining directions within the $X_{MT}Y_{MT}$ working plane.

Light Machine Tools for Productive Machining 103

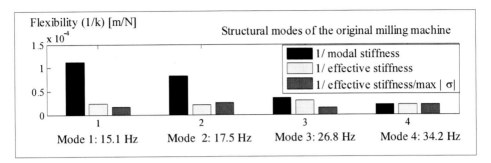

Figure 16. Additional analytical calculations on the FEM model of the original milling machine.

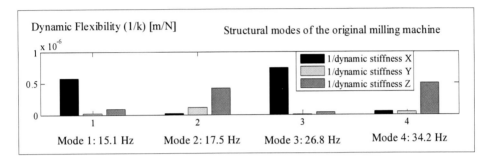

Figure 17. Dynamic stiffness associated to the original milling machine modes in Cartesian axes.

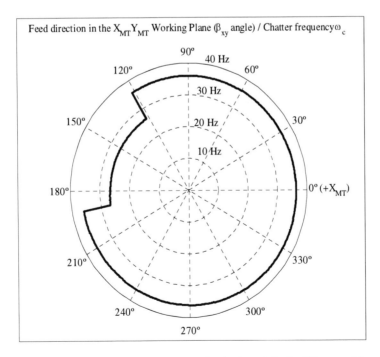

Figure 18. Polar diagram of structural chatter frequencies in the original milling machine. Spindle speed: 400=600 rpm.

As it can be seen in Figure 18, for the great majority of machining directions, the chatter frequency is in the range of 35 Hz (roughly speaking, in the first, the third and the fourth quadrants of the working plane) and in the range of 25 Hz (roughly speaking, in the second quadrant of the working plane). From a identification process among the chatter frequencies and the structural frequencies of the machine listed in Figure 15, it can be concluded that the chatter frequencies of this machine/process combination are associated with mode 3 involving vertical torsion of the column (at 26.8 Hz.) and mode 4 involving flexion of the ram and the frame on the vertical plane (at 34.2 Hz.). The fact that in some cases the chatter frequencies are below the natural frequencies and in other cases the chatter frequencies are above the natural frequencies is associated with the sign of the directional factor σ, as it can be concluded from Figure 7.

From this data, some conclusions of interest can be drawn:

i) If the mass and rigidity associated with these two limiting modes 3 and 4 is reduced, combined with a parallel increase in the same proportion of the damping associated with such modes, the productivity of the machine will be able to be maintained with using less material resources and with a reduction in energy consumption in the phase of use.

ii) If the mass and rigidity associated with non-limiting modes of the machine, such as modes 1 and 2, is reduced by continuously checking to ensure these changes do not alter the non-limiting nature of such modes, the productivity of the machine will be able to be maintained with using less material resources and with a reduction in energy consumption in the phase of use.

3.2. Re-design of the Horizontal Milling Machine

From the analytical and experimental study carried out on the horizontal milling machine, the following changes in mechanical design have been implemented in order to reduce the weight of the machine while at the same time maintaining and even increasing its productivity:

1) Lightening of the ram by using light materials and a high degree of internal damping: a ram has been designed based on steel sandwich and aluminium foam elements that reduces the total moving mass by 15% and increases internal damping by 250% in relation to a conventional steel ram, thus reducing static rigidity in directions X and Z by 10%. The ram guide systems have also been re-conceived in order to increase the damping associated with the ram flexion mode without this meaning an increase in the mass of the machine.
2) Lightening of the milling head: by replacing steel with cast aluminium, the weight is reduced by 27%, decreasing from the initial 550 kg to 400 kg without this affecting the productivity of the machine.
3) Reduction in conductive inertia: by reducing the inertia associated with the moving components and optimizing the dynamic behavior of the servo-drives, the total inertia reflected in the drive shaft has been able to be reduced by 50%, thus using servo-motors with less inertia in the servo-axes Y and Z.
4) Introduction of active damping systems: in order to improve the damping associated with the torsion mode of the column without modifying the column itself, an active vibration

control system has been incorporated into the horizontal ram based on an inertial actuator whose direction of application of force is the X axis.

The above changes have been introduced into the MEF model of the machine, which has enabled the effects of such changes to be virtually validated both with regard to the weight of the moving elements and the productivity of the machine - measured in terms of the amount of chip that the machine is capable of evacuating without there being instability.

3.3. Manufacture of a Prototype of a Productive and Eco-Efficient Milling Machine

Once the original horizontal milling machine has been re-designed in accordance with the criteria indicated in the previous subchapter, it can then be manufactured and assembled so as to be able to experimentally check the reductions in weight and increases in productivity when incorporating the different re-design techniques, in addition to the active vibration control system.

The mechanical and dynamic results that have been obtained regarding this prototype of productive and light milling machine, prior to carrying out consumption and productivity measuring tests, are as follows:

Table 3. Comparison of dynamic and mechatronic properties between the original milling machine and the Prototype of light and productive machine; data from Y axis

Y axis of the machine (longitudinal direction of the ram)			
Concept	Original machine	Prototype	Comparison
Maximum acceleration / continuous rating	1.3 / 3.1 m/s²	1.3 /2.9 m/s²	The dynamic functional nature of the machine is maintained
Moving mass	2,500 kg.	2,100 kg	16% reduction in the moving mass
Mass reflected in the motor	13,650 kg	7,085 kg	48% reduction in the mass reflected in the motor
Mechanical rigidity of the transmission	328 N/μm	219 N/μm	33% increase in the mechanical rigidity
Natural frequency in the drive	29 Hz	35 Hz	20% increase in the natural frequency
Bandwidth of the drive	6 Hz	7 Hz	21% increase in the mechatronic robustness
Output of the motor	10 kW	6 kW	40% reduction in the motor output

Table 4. Comparison of dynamic and mechatronic properties between the original milling machine and the Prototype of light and productive machine; data from Z axis

Z axis of the machine (vertical direction of the milling machine)			
Concept	Original machine	Prototype	Comparison
Maximum acceleration / continuous rating	1.3 / 2.1 m/s²	1.3 /1.9 m/s²	The dynamic functional nature of the machine is maintained
Moving mass	4,500 kg.	3,700 kg	18% reduction in the moving mass
Mass reflected in the motor	15,415 kg	8,965 kg	42% reduction in the mass reflected in the motor
Mechanical rigidity of the transmission	430 N/μm	300 N/μm	30% reduction in the mechanical rigidity
Natural frequency in the drive	26 Hz	30 Hz	15% reduction of the natural frequency
Band width of the drive	5 Hz	6 Hz	17% increase in the mechatronic robustness
Output of the motor	10 kW	6 kW	40% reduction in the motor output

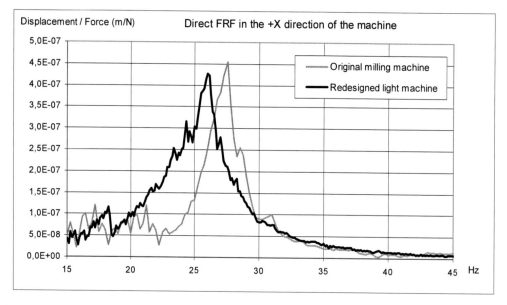

Figure 19. Comparison between direct FRFs in direction X: original milling machine vs. Prototype of light-weight machine.

Below the *Frequency Response Functions* (FRF) are shown, which have been measured in the prototype of a productive and light-weight machine, in addition to their comparison with the original milling machine with similar functions. As a starting point, Figure 19 shows a comparison between FRFs in direction X of the machine.

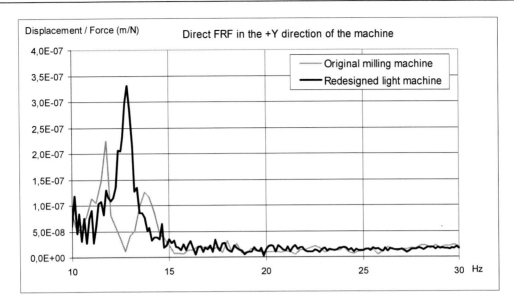

Figure 20. Comparison between direct FRFs in direction Y: original milling machine vs. Prototype of light-weight machine.

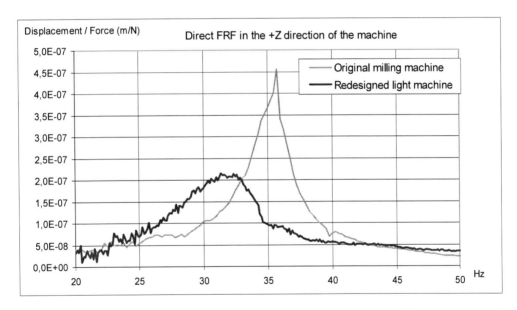

Figure 21. Comparison between direct FRFs in direction Z: original milling machine vs. Prototype of light-weight machine.

By comparing both FRFs, it can be seen how the structural mode of the initial machine, associated with torsion of the column, has been transformed into a mode with more passive damping and with greater dynamic flexibility which has remained practically invariable. Both effects are positive from the eco-efficiency point of view, as on the one hand, the Prototype of light and productive machine has less mass (the ram weighs 15% less) which does not yet jeopardize productivity due to an increase in damping. On the other hand, it should be taken into account that no active vibration control system has been included in these FRFs – an aspect that will result in an additional increase in productivity at the expense of a minimal increase in energy consumption.

Furthermore, Figure 20 shows another comparison of direct FRFs between the re-designed machine and the original milling machine, in this case, for direction Y of the machine (horizontal feed of the ram).

In this case, the Prototype evidences less dynamic rigidity than in the initial machine. This does not constitute any problem: on the contrary, it has been the result of a deliberate strategy, as the stability lobe diagrams indicated that the machine modes on the Y axis did not limit the productivity of the machine. This has enabled the ram to be re-designed with less dynamic rigidity (and therefore with less weight) without affecting the productivity of the machine. Lastly, Figure 21 shows the comparison of FRFs in this case for direction Z (vertical movement of the frame).

The most noteworthy aspect of this comparison of FRFs is the high degree of passive damping that has been obtained for the structural flexion mode of the ram on the plane YZ. This aspect is of great importance because the initial milling machine evidenced at a natural frequency of 36 Hz, a high modal and effective stiffness (and therefore, a high energy consumption), whereas the light Prototype evidences at a natural frequency of 32 Hz, a lower modal stiffness, although, owing to the effect of the damping, it manages to reduce the dynamic flexibility from the initial 4.44e-7 m/N to the final 2.15e-7 m/N. All this translates into the fact that the Prototype increases its eco-efficiency in the vertical direction (less mass and less energy consumption) while at the same time increases its productivity (higher dynamic stiffness) measured in terms of *material removal rate*.

3.4. Introduction of Active Damping in the Prototype of Light-Weight Machine

Figure 22 shows the active damping device based on an inertial actuator embedded in the horizontal ram of the Prototype. This has been possible due to re-designing the ram in order to make it lighter, and designing a housing into which the active damping device can be placed without jeopardizing the functionalities of the machine: useful courses, collisions, etc.

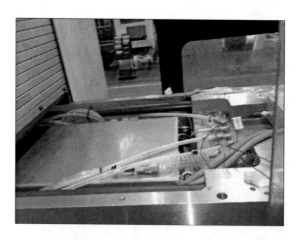

Figure 22. Active Damping Device embedded in the ram of the light-weighted Prototype.

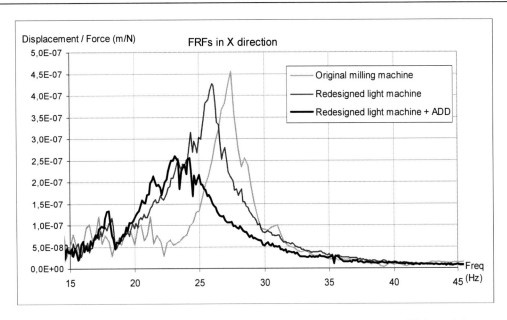

Figure 23. Direct FRFs in X_{MT} direction: Original milling machine vs. Prototype of light-weight machine, with and without ADD.

When characterizing the damping effect of these inertial actuators, the procedure that has been followed is that of measuring combinations of *Frequency Response Functions* (FRFs) associated with different excitation and response directions.

As a reference, Figure 23 shows the additional modal damping that has been obtained in the longitudinal direction of the bench (X+ axis) when the torsion mode of the column is damped with the inertial actuator. The FRF in Figure 23 shows the effect of the damper on the dynamic stiffness of the machine at the TCP.

Figure 23 is of great interest, as it shows damping at 40% of the dynamic flexibility associated with mode 3 of the machine, which is equivalent to an increase of 67% of the relative damping coefficient ξ associated with that mode in relation to the initial coefficient. This means that the critical depth of cut for machining processes in which the limiting mode with regard to the appearance of self-induced vibrations in mode 3 will be around 67% higher than the initial depth.

Taking into account the fact that the output for the active vibration control system is 200 watts (around 0.5% of installed power in the original milling machine), this active damping system enables a high material removal rate without affecting the energy consumption of the machine. Later, as this machine has been re-designed in order to reduce the weight of its moving components, the productivity of the machine can be increased while at the same time consuming less energy while it carries out machining operations.

Once the new FRFs have been measured at the TCP of the Prototype, the stability lobe diagrams have then been calculated for a milling process with a tool of 125 mm in diameter and nine teeth, machined in different directions within the XY work plane of the machine. The diagrams have been calculated initially with the active damper deactivated, and subsequently with the damper activated.

Figure 24 shows eight different stability lobe diagrams, which correspond to gradual variations in the machining direction at 45°. The maximum depth of cut is shown for each machining direction for different rotational speeds of the tool, and also for the cases of the activated and deactivated vibration control system.

On the other hand, for the machining direction at 180 ° (direction X- of the machine), and focusing on the range of working speeds of the tool that is being taken into consideration – i.e., from 300 to 700 rpm – the graphs are obtained which are shown in Figure 25. For this case, the lobe diagram on the left shows increases higher than 100% in terms of maximum depth of cut values – an improvement that is associated with the increase in modal damping in structural modes that limit the productivity of the machine.

In the diagram on the right, it can also be seen how the active vibration control associated with the structural mode that limits the productivity of the machine, which is in a lower frequency mode, gives rise to a jump in the vibration frequency. This means that the initial bending mode (i.e. mode 4) stops being limiting and a mode of a higher frequency becomes limiting, with significantly greater modal rigidity.

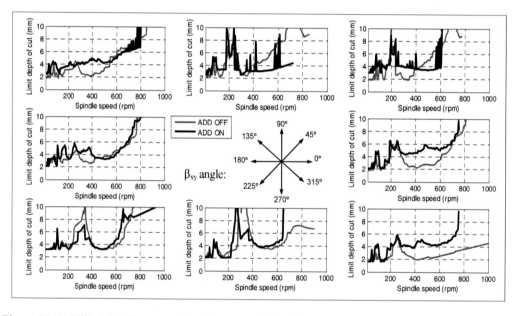

Figure 24. Stability lobe diagrams of the Prototype with and without active vibration control for different machining directions (β_{xy} angle) in the working plane $X_{MT}Y_{MT}$.

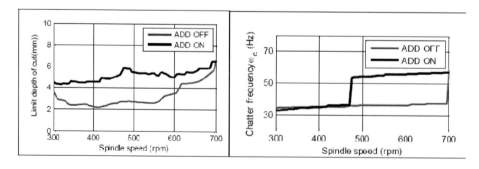

Figure 25. Zoom of the active damping in direction X=180° (X-) of the Prototype; rotational speeds of 300 – 700 rpm.

4. VALIDATION OF THE PROTOTYPE OF PRODUCTIVE AND ECO-EFFICIENT MILLING MACHINE

4.1. Tests for Measuring Productivity: Material Removal Rate MRR

The active damping effect has proved especially noticeable in the direction of actuation, which in the Prototype ram has been direction ± X. The increase in modal damping in the bandwidth of actuation of the Prototype has enabled the dynamic flexibility associated with the milling machine to be reduced and thus obtain an increase in the critical depths of cut.

Figure 26 shows the effect active damping has meant in the different machining directions within the $X_{MT}Y_{MT}$ work plane of the Prototype. This Figure shows a polar diagram that relates machining directions and maximum axial depths of cut that prevent the appearance of instability in the machining process. 0° coincides with the $+X_{MT}$ direction of the machine.

Figure 26 shows how in machining direction $+X_{MT}$ (0° angle in the polar diagram), the axial critical depth of cut of the Prototype has been increased by over 100% when activating the active control of vibrations. This is equivalent to an increase in the productivity of the machine in MRR to the same extent – that is, by 100% - when the increase in energy consumption is lower, as the inertial actuator of which the vibration control system operates at a maximum output of 200 watts, i.e. around 0.5% of the total output of the Prototype, and the mobile structural elements involved in the $+X_{MT}$ motion weigh approximately 20% less.

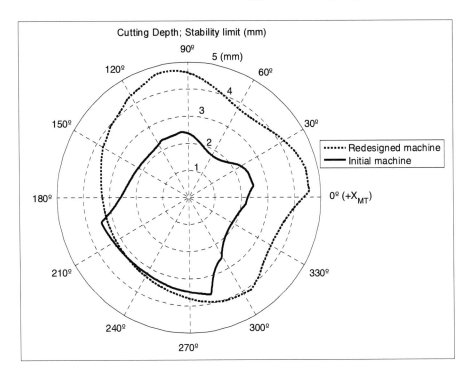

Figure 26. Productivity of the Prototype of light and productive machine in different machining directions.

This increase in productivity is not confined to the direction of the actuator. Figure 26 indicates that in machining direction Y+ (90° in the polar diagram), the maximum critical depth of cut is also increased by 100%, with the vibration control system activated – in other words, in machining direction Y+, the Prototype has obtained a further 100% increase in the productivity of the machining processes.

If it is taken into account that the Prototype has obtained an increase in productivity of 100% in directions +X y +Y, and that this increase in productivity has been obtained in a milling machine that consumes less energy than a conventional milling machine of the same functional nature, the conclusion can be drawn that the Prototype combines an increase in productivity and an increase in energy efficiency in such a way that the Prototype validates a new concept in the machine tool sector: eco-productivity.

In the following section, the reduction in energy consumption of the Prototype will be stressed in more detail and in a more quantified way.

4.2. Tests to Measure Energy Consumption

4.2.1. No-Load Tests

The lightened features of the Prototype, with reduction in weight of over 40% in the inertias reflected in drives Y and Z, mean that its servo-mechanisms reach high feed speeds and high rates of acceleration while at the same time consuming less energy. Expressed in a quantified way, Figure 27 shows a comparison between the energy consumption associated with the re-designed light-weight machine and a conventional milling machine of the same dimensions and functions when both milling machines are subjected to the same positioning references and in both cases, acceleration is the maximum that can be obtained by their servo-mechanisms.

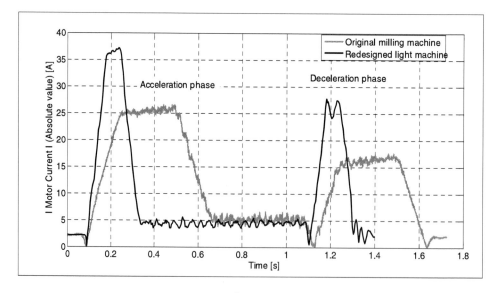

Figure 27. Comparison of energy consumption between axis-Y motors in the Prototype of light and productive milling machine and the original milling machine.

Specifically, Figure 27 shows the temporary movement profiles that are associated with the positioning of the tool at 30 m/min and at the maximum acceleration the servo-mechanism is able to reach. The continuous line shows this positioning process for the original milling machine, whereas the dotted line shows the same case of positioning, but in this case for the Prototype.

Integration over time of the power consumed by the motors of the Y servo-mechanism in both cases indicates that the conventional milling machine consumes 31.9 joules per each positioning cycle, whereas commanding the same profile of movement, the Prototype consumes 21.04 joules during the same cycle. This means that the Prototype of eco-efficient milling machine maintains and even increases productivity in terms of *material removal rate* while at the same time consuming 35% less energy during no-load movements of the tool.

4.2.2. Machining Tests

Below some machining tests that have been carried out with the Prototype of a productive and eco-efficient milling machine are shown. Measurements have been taken for certain AISIS 1045 steel roughing operations with a plate of 125 mm, 9 flutes at a 45° angle and with 80% radial immersion in concordance, rotation of the milling head at 400 rpm and 720 mm/min feed speed – that is, with a feed per tooth of 0.2 mm. The tests have been carried out for different depths of cut and with the vibration damping system activated and deactivated.

Thus, Figure 28 shows the active power consumed by the milling head in certain roughing operations in which the vibration damping system has been activated in some sections.

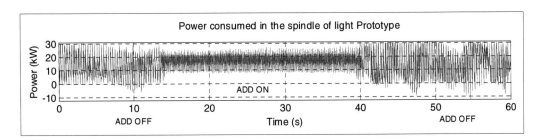

Figure 28. Active power consumption in the head of the Prototype, with and without ADD.

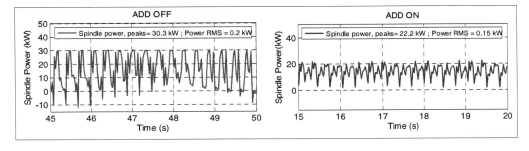

Figure 29. Power consumed by the milling head with ADD activated/deactivated.

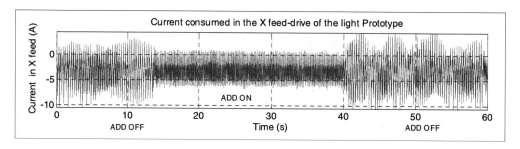

Figure 30. Comparison of consumption of current in drive X with ADD activated/deactivated.

For its part, Figure 29 shows the data from Figure 28 with greater resolution on the axis of abscissa (time axis). In this figure, it can be seen that active damping of vibrations reduces both the peaks and the average consumption of active power in the milling head.

By integrating the above power consumption levels in the domain of time, a comparison can be made between the energy consumption in kW·h at intervals of activated and deactivated active damping of vibrations. For machining intervals of 10 seconds, energy consumption with control deactivated has been 0.043 Kwh, whereas with the active vibration damping system on, consumption for the same interval has been 0.040 Kwh - i.e. a reduction of 7% has been obtained in the consumption of electrical energy.

On the other hand, active vibration control also reduces the energy consumed by the servo-drive in the feed direction. Thus, Figure 30 shows a comparison between the instantaneous current consumed by the servo-drive motor of the X axis with the vibration damping system activated and deactivated.

Expressed in a quantified way, the reduction in consumption peaks has been 37%, whereas the average power consumed by the drive has been reduced by 17%.

5. CONCLUSION

This chapter has introduced a design methodology aimed at conceiving milling machines in which high productivity is combined with reduced environmental impact – an aspect that is related to a large extent with the consumption of material resources in the construction phase and energy consumption in the phase of use. The approach has been based on breaking the link between a stable machine-part interaction during the machining process and the need to obtain high levels of mechanical rigidity by means of structural components of great weight.

The construction of a prototype of a eco-efficient and productive milling machine, in which reductions of up to 50% have been obtained in the inertias reflected in the drive shafts in relation to the conventional milling machine, has enabled reductions of 35% to be obtained in the energy consumed in non-load movements in relation to the milling machine referred to above. Parallel to this, increases in the *material removal rate* of 100% have been obtained in some machining directions. On the other hand, the integration of an active vibration control system has enabled reductions in energy consumption to be obtained both in the milling head itself and in the feed servo-drives in terms of the energy consumption of the Prototype, without vibration damping.

All of this leads us to draw the conclusion that a multidisciplinary approach aimed at maximizing the ratio between the mass involved in a machine tool and the productivity level reached with that machine has a beneficial effect both on the machine supplier and the its user as, on one hand, manufacturing and machine purchase costs are reduced and, on the other, operating costs are reduced. Indeed, this mutual benefit, combined with a modular conception of the machines that enables light and productive machines to be put together swiftly, may in turn, enable new business models to be established in the manufacturing sector among suppliers and users of machine tools, such as sharing benefits, risks and costs associated with manufacturing processes.

REFERENCES

[1] Dietmair A. et al. (2010), Lifecycle Impact Reduction and Energy Savings through Lightweight Eco-Design of Machine Tools. 17th CIRP International Conference on LCE, Anhui, China.

[2] Henninger C., Eberhard P. (2007), An Investigation of Pose-Dependent Regenerative Chatter for a Parallel Kinematic Milling Machine. Proceedings of the IFToMM 12th World Congress in Mechanism and Machine Science, Besançon, France.

[3] Bravo U. (2007), Un procedimiento para la predicción de la estabilidad dinámica en el mecanizado a alta velocidad de paredes delgadas (in Spanish). Doctoral Thesis in the University of the Basque Country UPV/EHU.

[4] Dequidt A. et al. (2000), Mechanical pre-design of high performance motion servo-mechanisms. *Mechanism and Machine Theory* 35 pp. 1047-1063.

[5] López de Lacalle, L.N.; Lamikiz, A. (Eds.) (2009), Machine Tools for High Performance Machining, pp. 1-44.

[6] Zulaika J. et al. (2010), Eco-efficient and highly productive production machines by means of a holistic Eco-Design approach. Proceedings of the E/E 3rd International Conference on Eco-Efficiency. Egmond aan Zee, Netherlands.

[7] Zulaika J. Campa, F.J. (2008), Optimización de los parámetros de diseño de una máquina-herramienta en base a criterios de productividad y ecoeficiencia (in Spanish), XVII Congress on Machine Tools and Manufacturing Technologies, San Sebastián, Spain.

[8] Altintas, Y. (2001), Analytical Prediction of Three Dimensional Chatter Stability in Milling, JSME Int. J. Series C: Mechanical Systems, Machine Elements and Manufacturing, 44, n.3.

[9] Zatarain, M., Muñoa, J., Peigné, G., Insperger, T. (2006), Analysis of the Influence of Mill Helix Angle on Chatter Stability, Annals of CIRP, Vol. 55, No. 1, pp 365-368.

[10] Zulaika J. et al. (2010), Highly productive machining processes and eco-efficient machine tools by means of an integrated machine+process approach, Proceeding of the 4th International Conference of High Speed Machining ICHMS2010. Guangzhou (China).

[11] Altintas, Y., Budak, E. (1995), Analytical Prediction of Stability Lobes in Milling, Annals of the CIRP, 44/ 1: 357-362.

[12] Zatarain, M., Insperger, T., Peigne, G., Villasante, C., Muñoa, J. (2007), Analysis of Directional Factors in Milling: Importance of Multifrequency Calculation and of the Inclusion of the Effect of the Helix Angle, 6th Int. Conf. on High Speed Machining, San Sebastian, Spain.

[13] Muñoa, J., et al (2005), Optimization of Hard Material Roughing by Means of a Stability Model, 8th CIRP Int. Workshop: Modelling of Machining Operations, Chemnitz.

[14] Zulaika J. et al (2009), Stability Lobe Diagrams for the Re-design of a Machine-Tool based on Ecoefficiency Criteria, 12th CIRP Conf. on Model. of Mach Operations, San Sebastian.

[15] Altintas, Y., (2000), Manufacturing Automation: Metal Cutting Mechanics, Machine Tool Vibrations, and CNC Design; Cambridge University Press.
[16] Ganguli, A. Deraemaeker, A. Preumont, A. (2007), Regenerative chatter reduction by active damping control, *Journal of sound and vibration*, vol. 300, pp. 847-862.

In: Machine Tools: Design, Reliability and Safety
Editor: Scott P. Anderson, pp. 117-151

ISBN: 978-1-61209-144-0
© 2011 Nova Science Publishers, Inc.

Chapter 4

IMPROVING MACHINE TOOL PERFORMANCE THROUGH STRUCTURAL AND PROCESS DYNAMICS MODELING

Tony L. Schmitz, Jaydeep Karandikar, Raul Zapata, Uttara Kumar, and Mathew Johnson
University of Florida, Machine Tool Research Center,
Department of Mechanical and Aerospace Engineering,
Gainesville, FL, USA

ABSTRACT

There are many factors that influence the performance of computer-numerically controlled machining centers. These include tool wear, positioning errors of the tool relative to the part, spindle error motions, fixturing concerns, programming challenges, and the machining process dynamics. In this study, the limitations imposed by the process dynamics are considered. Algorithms used to predict the tool point frequency response function and, subsequently, the stability and surface location error (due to forced vibrations) are described. This information is presented graphically in the form of the milling "super diagram", which also includes the effect of tool wear and incorporates uncertainty in the form of user-defined safety margins. Given this information at the process planning stage, the programmer can select optimized operating parameters that increase the likelihood of first part correct production and reduce the probability of damage to the tool, spindle, and/or part due to excessive forces and deflections.

1. INTRODUCTION

Critical elements of a modern computer-numerically controlled (CNC) machining center include the: 1) computer-aided design and manufacturing (CAD/CAM) software; 2) machinist; 3) human-machine interface; 4) control system; 5) machine structure; 6) positioning system (drives, guideways, table); 7) part fixture(s); and 8) spindle-holder-tool assembly. Machine tool designers couple their experience with computer-aided modeling tools and expend significant effort in applying these capabilities to produce machining centers that meet customer requirements for accuracy, throughput, reliability, and safety. However, the selection of the holder and cutting tool is generally left to the customer based on the particular application.

While this paradigm is reasonable given the broad range of potential part geometry and material combinations that may be implemented on a particular CNC machine, it places the user in a difficult position because the cutting tool is often the most flexible element in the machine tool structure. Its compliance, therefore, can limit productivity due to forced and self-excited (chatter) vibrations during material removal. This chapter describes recent efforts to predict the machine-spindle-holder-tool dynamics and use this information to select preferred combinations of spindle speeds and depths of cut at the process planning stage for profit maximization.

Using a procedure referred to as Receptance Coupling Substructure Analysis (RCSA), the machine-spindle dynamics are coupled to dynamic models of the holder-tool to predict the assembly's dynamic response (frequency response function). Given this information, frequency-domain algorithms may be applied to select the spindle speed-depth of cut pair that maximizes profit for a selected part. This physics-based, pre-process selection of operating parameters reduces the time to production and improves system reliability by minimizing the occurrence of chatter and large vibrations which can damage the tool, part, and spindle.

2. BACKGROUND

Limitations to milling productivity include tool wear, positioning errors of the tool relative to the part, spindle error motions, fixturing concerns, programming challenges, and the process dynamics. Many research studies have been completed to address these issues individually. Tobias, Tlusty, and Merrit defined the mechanism for self-excited vibrations (chatter) as *regeneration of waviness* [1-3], where the current chip thickness depends on the commanded chip thickness, the surface left by the previous tooth (for milling), and the current vibration state of the cutter. This fundamental understanding led to the development of the stability lobe diagram, which represents stable and unstable combinations of spindle speed and axial depth of cut for milling in a graphical format. Frequency-domain solutions were proposed by Tlusty *et al.* [4] and Altintas and Budak [5]. A representation of a typical stability lobe diagram is provided in Figure 1. To produce the diagram, the following information is required:

- the dynamic response of the tool-holder-spindle-machine as reflected at the free end of the endmill (i.e., the tool point); the part dynamics may also be considered
- the cutting force model, which relates the cutting force components to the uncut chip area; this model may be established mechanistically [6-7] or through the use of orthogonal cutting mechanics to represent oblique cutting cases [6]
- basic cutting information including the radial depth of cut, cutting direction (relative to the measured or modeled tool point dynamics), and number of teeth on the cutter.

In additional to chatter, the process dynamics can limit milling performance even under stable cutting conditions. In this case, the forced vibrations caused by the constant entrance and exit of the teeth from the part can lead to an error in the location of the machined surface, referred to as surface location error (SLE) [8-19]. Depending on the relationship between the selected spindle speed (or tooth passing frequency) and the system natural frequency which corresponds to the dominant tool point mode, more material may be removed than commanded (resulting in an overcut surface) or less material may be removed (undercut). A case study presented in [20] showed that SLE can be the dominant error source for typical milling operations on a modern CNC machining center. A frequency-domain algorithm for SLE prediction was developed by Schmitz and Mann [21] to complement the frequency-domain stability solutions. Similar to the stability algorithms, the SLE prediction algorithm requires the tool point frequency response, force model, radial depth, cutting direction, and number of teeth. Additionally, the tool helix angle must be known due variation in SLE along the tool axis for helical endmills [22].

Tool wear has also naturally received significant attention in the literature. Taylor first established an empirical basis for the relationships between cutting parameters and tool wear in 1906 [23] and many subsequent studies have been completed to extend this work. In particular, tool condition monitoring systems have recently been implemented to estimate tool wear based on various transducer signals [24-25]. Other issues related to limitations on milling productivity are detailed in relevant textbooks; see [6-7, 22, 26-28], for example.

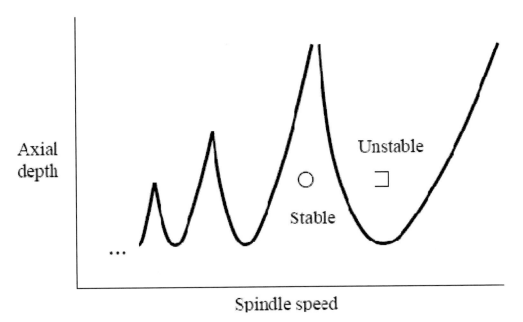

Figure 1. Typical stability lobe diagram. The stability boundary separates stable spindle speed-axial depth combinations (below the boundary, marked as a circle) from unstable pairs (above, marked as a square).

In order to simultaneously consider the productivity limits imposed by both the process dynamics and tool wear, the milling "super diagram" has been proposed [29]. This diagram combines stability and SLE information in a graphical format using a gray-scale color coding scheme. The deterministic stability limit [5] and SLE values [21] for a selected spindle speed-axial depth of cut domain are calculated based on the system dynamic response, cutting force model, and other process/tool information (as noted previously). The effects of tool wear may be incorporated by applying wear-dependent cutting force coefficients [30] to the calculation of the stability limit and SLE values. By correlating the tool wear status with the change in force coefficient values, the process dynamics can be tailored to the behavior of a new tool or one at or near its end of life. Additionally, model and input data uncertainties may be included as a user-specified safety margin. The safety margin captures the user's beliefs regarding how close (in spindle speed and axial depth) he/she is willing to operate relative to the deterministic limits obtained using the frequency-domain models.

3. RECEPTANCE COUPLING SUBSTRUCTURE ANALYSIS

As described in Section 2, in order to predict the milling dynamics, the tool point frequency response function (FRF) must be known. Experimental techniques are available to measure this FRF. Typically, impact testing is implemented where an instrumented hammer is used to excite the tool point (with a short duration impulse, or impact) and a linear transducer, such as a low mass accelerometer, is used to measure the response [7, 31]. This provides a convenient approach to obtaining the tool point FRF, but requires a separate set of measurements for each tool-holder-spindle-machine assembly. This can be problematic in production facilities, where there can be several hundred combinations, due to the significant measurement time.

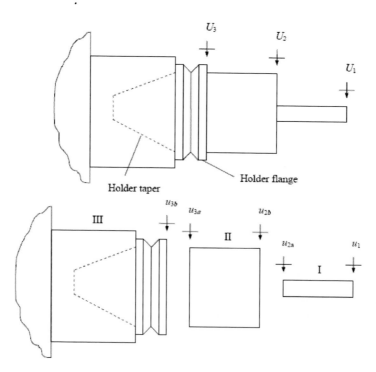

Figure 2. Three-component receptance coupling model of tool (I), holder (II), and spindle-machine (III).

As an alternative to measuring each tool-holder-spindle-machine combination, this assembly can be considered as three separate components: the tool, holder, and spindle-machine and the individual frequency response of these components can be coupled; see Figure 2. Of these three, the tool and holder are convenient to model, while the spindle-machine, on the other hand, is not as well-suited to modeling. Spindle dynamics modeling, often completed using finite element analysis, requires detailed knowledge of the mechanical design, bearing stiffness values, and damping levels. This information is often unavailable to the machine user. Similarly, adequate information regarding the machine structure, especially the appropriate damping values at joints, is typically difficult to obtain without performing separate measurements.

In the RCSA approach, the tool and holder FRFs, or receptances, are determined from simple beam models and the spindle-machine receptances are measured by impact testing. These substructure receptances are then joined analytically to obtain the assembly FRF as reflected at the tool point [32-44]. The three-component RCSA model is reviewed in the following paragraphs [38, 7].

Figure 2 depicts the three individual components of the tool-holder-spindle-machine assembly: I – tool, II – holder, and III – spindle-machine. Both the tool and holder are described using Timoshenko beam models (based on the geometry and material properties) with free-free boundary conditions [45]. The free-free tool and holder models are then coupled to form the subassembly I-II identified in Figure 3, where $u_i = \begin{Bmatrix} x_i \\ \theta_i \end{Bmatrix}$ are the component generalized coordinates composed of both displacement, x_i, and rotation, θ_i and $U_i = \begin{Bmatrix} X_i \\ \Theta_i \end{Bmatrix}$ are the assembly generalized coordinates. To couple components I and II, the coordinate definitions provided in Figure 3 are applied, where $q_i = \begin{Bmatrix} f_i \\ m_i \end{Bmatrix}$ are the component generalized forces composed of both force, f_i, and moment, m_i, and $Q_1 = \begin{Bmatrix} F_1 \\ M_1 \end{Bmatrix}$ is the assembly generalized force applied at assembly coordinate 1. The component I receptances include: 1) the direct receptances at the free end $h_{11} = \dfrac{x_1}{f_1}$, $l_{11} = \dfrac{x_1}{m_1}$, $n_{11} = \dfrac{\theta_1}{f_1}$, and $p_{11} = \dfrac{\theta_1}{m_1}$; 2) the cross receptances from the free end to the fixed end (connected to the holder) $h_{12a} = \dfrac{x_1}{f_{2a}}$, $l_{12a} = \dfrac{x_1}{m_{2a}}$, $n_{12a} = \dfrac{\theta_1}{f_{2a}}$, and $p_{12a} = \dfrac{\theta_1}{m_{2a}}$; 3) the direct receptances at the fixed end $h_{2a2a} = \dfrac{x_{2a}}{f_{2a}}$, $l_{2a2a} = \dfrac{x_{2a}}{m_{2a}}$, $n_{2a2a} = \dfrac{\theta_{2a}}{f_{2a}}$, and $p_{2a2a} = \dfrac{\theta_{2a}}{m_{2a}}$; and 4) the cross receptances from the fixed end to the free end $h_{2a1} = \dfrac{x_{2a}}{f_1}$, $l_{2a1} = \dfrac{x_{2a}}{m_1}$, $n_{2a1} = \dfrac{\theta_{2a}}{f_1}$, and $p_{2a1} = \dfrac{\theta_{2a}}{m_1}$. These are organized into the generalized component receptance matrices, $R_{ij} = \begin{bmatrix} h_{ij} & l_{ij} \\ n_{ij} & p_{ij} \end{bmatrix}$, where $\{u_i\} = [R_{ij}]\{q_j\}$. For component II, the same receptances must be calculated using the Timoshenko beam model, but coordinate 1 is replaced with 2b and coordinate 2a is replaced with 3a. By assuming a rigid coupling between these two components, the I-II subassembly tip receptances: (direct) G_{11} and G_{3a3a}; and (cross) G_{13a} and G_{3a1} can be determined, where $G_{ij} = \begin{bmatrix} H_{ij} & L_{ij} \\ N_{ij} & P_{ij} \end{bmatrix}$ are the generalized assembly receptance matrices and $\{U_i\} = [G_{ij}]\{Q_j\}$.

To determine the direct and cross receptances at the right end of the subassembly, G_{11} and G_{3a1}, Q_1 is applied to coordinate U_1 as shown in Figure 3. The components' displacements/rotations are: $u_1 = R_{11}q_1 + R_{12a}q_{2a}$, $u_{2a} = R_{2a1}q_1 + R_{2a2a}q_{2a}$, $u_{2b} = R_{2b2b}q_{2b}$, and $u_{3a} = R_{3a2b}q_{2b}$. The equilibrium conditions are: $q_{2a} + q_{2b} = 0$ and $q_1 = Q_1$. The component displacements/rotations and equilibrium conditions are substituted into the compatibility condition for the rigid connection, $u_{2b} - u_{2a} = 0$, to obtain the expression for q_{2b} shown in Eq. 1. The component force q_{2a} is then determined from the equilibrium condition $q_{2a} = -q_{2b}$. The expression for G_{11} is provided in Eq. 2. The cross receptance matrix G_{3a1} is shown in Eq. 3.

$$\begin{aligned}
u_{2b} - u_{2a} &= 0 \\
R_{2b2b}q_{2b} - R_{2a1}q_1 - R_{2a2a}q_{2a} &= 0 \\
(R_{2a2a} + R_{2b2b})q_{2b} - R_{2a1}Q_1 &= 0 \\
q_{2b} &= (R_{2a2a} + R_{2b2b})^{-1} R_{2a1} Q_1
\end{aligned} \qquad (1)$$

$$G_{11} = \dfrac{U_1}{Q_1} = \dfrac{u_1}{Q_1} = \dfrac{R_{11}q_1 + R_{12a}q_{2a}}{Q_1} = \dfrac{R_{11}Q_1 - R_{12a}(R_{2a2a} + R_{2b2b})^{-1} R_{2a1} Q_1}{Q_1}$$

$$G_{11} = R_{11} - R_{12a}(R_{2a2a} + R_{2b2b})^{-1} R_{2a1} = \begin{bmatrix} H_{11} & L_{11} \\ N_{11} & P_{11} \end{bmatrix} \qquad (2)$$

$$G_{3a1} = \dfrac{U_{3a}}{Q_1} = \dfrac{u_{3a}}{Q_1} = \dfrac{R_{3a2b}q_{2b}}{Q_1} = \dfrac{R_{3a2b}(R_{2a2a} + R_{2b2b})^{-1} R_{2a1} Q_1}{Q_1}$$

$$G_{3a1} = R_{3a2b}(R_{2a2a} + R_{2b2b})^{-1} R_{2a1} = \begin{bmatrix} H_{3a1} & L_{3a1} \\ N_{3a1} & P_{3a1} \end{bmatrix} \qquad (3)$$

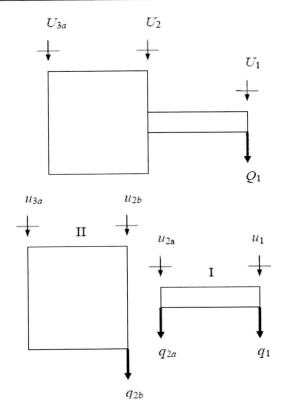

Figure 3. Subassembly I-II composed of tool (I) and holder (II). The generalized force Q_1 is applied to U_1 to determine G_{11} and G_{3a1}.

To find the remaining tip receptances G_{3a3a} and G_{13a}, Q_{3a} is applied to coordinate U_{3a}. Following the same approach, the equations for the direct receptance G_{3a3a} (Eq. 4) and the cross receptance G_{13a} (Eq. 5) are determined.

$$G_{3a3a} = \frac{U_{3a}}{Q_{3a}} = \frac{u_{3a}}{Q_{3a}} = \frac{R_{3a3a}q_{3a} + R_{3a2b}q_{2b}}{Q_{3a}}$$

$$G_{3a3a} = \frac{R_{3a3a}Q_{3a} - R_{3a2b}(R_{2a2a} + R_{2b2b})^{-1}R_{2b3a}Q_{3a}}{Q_{3a}} \quad (4)$$

$$G_{3a3a} = R_{3a3a} - R_{3a2b}(R_{2a2a} + R_{2b2b})^{-1}R_{2b3a} = \begin{bmatrix} H_{3a3a} & L_{3a3a} \\ N_{3a3a} & P_{3a3a} \end{bmatrix}$$

$$G_{13a} = \frac{U_1}{Q_{3a}} = \frac{u_1}{Q_{3a}} = \frac{R_{12a}q_{2a}}{Q_{3a}} = \frac{R_{12a}(R_{2a2a} + R_{2b2b})^{-1}R_{2b3a}Q_{3a}}{Q_{3a}} \quad (5)$$

$$G_{13a} = R_{12a}(R_{2a2a} + R_{2b2b})^{-1}R_{2b3a} = \begin{bmatrix} H_{13a} & L_{13a} \\ N_{13a} & P_{13a} \end{bmatrix}$$

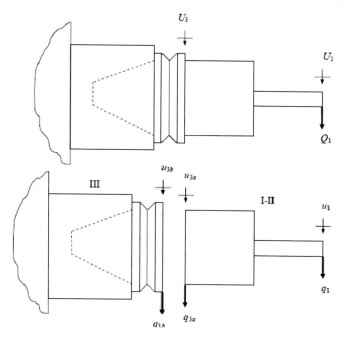

Figure 4. The I-II subassembly is rigidly coupled to the spindle-machine (III) to determine the tool point receptances, G_{11}.

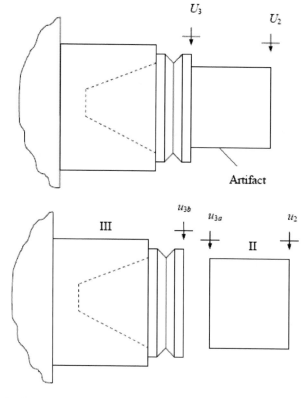

Figure 5. Artifact model for determining R_{3b3b} by inverse RCSA.

Once the free-free components I and II are coupled to form the subassembly I-II, this subassembly is then rigidly coupled to the spindle-machine to give the assembly tool point receptances, G_{11}; see Figure 4. This coupling is carried out using Eq. 6,

$$G_{11} = R_{11} - R_{13a}(R_{3a3a} + R_{3b3b})^{-1} R_{3a1}, \qquad (6)$$

where the R_{ij} matrices are the subassembly matrices from the I-II coupling result. Therefore, $R_{11} = G_{11}$ from Eq. 2, $R_{3a1} = G_{3a1}$ from Eq. 3, $R_{3a3a} = G_{3a3a}$ from Eq. 4, and $R_{13a} = G_{13a}$ from Eq. 5. The remaining unknown in Eq. 6 is the spindle-machine receptances R_{3b3b}.

In order to identify R_{3b3b} experimentally, a measurement artifact that includes not only the flange and taper, but also incorporates some length beyond the flange as shown in Figure 5 is inserted into the spindle. The assembly matrix $G_{22} = \begin{bmatrix} H_{22} & L_{22} \\ N_{22} & P_{22} \end{bmatrix}$ is determined experimentally and then the portion of the artifact beyond the flange is removed in simulation to isolate R_{3b3b}. The free end response for the artifact-spindle-machine assembly is provided in Eq. 7, where the R_{22}, R_{23a}, R_{3a3a}, and R_{3a2} matrices are populated using a beam model of the portion of the artifact beyond the flange. Equation 7 is rearranged in Eq. 8 to isolate R_{3b3b}. This step of decomposing the measured assembly receptances, G_{22}, into the modeled substructure receptances, R_{3a2}, R_{22}, R_{23a}, and R_{3a3a}, and spindle-machine receptances, R_{3b3b}, is referred to as "inverse RCSA".

$$G_{22} = R_{22} - R_{23a}(R_{3a3a} + R_{3b3b})^{-1} R_{3a2} \qquad (7)$$

$$\begin{aligned} G_{22} - R_{22} &= -R_{23a}(R_{3a3a} + R_{3b3b})^{-1} R_{3a2} \\ R_{23a}^{-1}(R_{22} - G_{22})R_{3a2}^{-1} &= (R_{3a3a} + R_{3b3b})^{-1} \\ R_{3a2}(R_{22} - G_{22})^{-1} R_{23a} &= R_{3a3a} + R_{3b3b} \\ R_{3b3b} &= R_{3a2}(R_{22} - G_{22})^{-1} R_{23a} - R_{3a3a} \end{aligned} \qquad (8)$$

In order to define the G_{22} receptances using the artifact, the direct displacement-to-force term $H_{22} = \dfrac{X_2}{F_2}$ is measured by impact testing. To find the rotation-to-force receptance $N_{22} = \dfrac{\Theta_2}{F_2}$, a first-order finite difference approach [46] is implemented. By measuring both the direct FRF H_{22} and cross FRF $H_{2a2} = \dfrac{X_{2a}}{F_2}$, N_{22} is computed using Eq. 9. The displacement-to-force cross FRF H_{2a2} is obtained by exciting the assembly at U_2 and measuring the response at coordinate U_{2a}, located a distance S from the artifact's free end, as shown in Figure 6.

$$N_{22} = \frac{H_{22} - H_{2a2}}{S} = \frac{H_{22} - H_{22a}}{S} \qquad (9)$$

By assuming reciprocity, L_{22} can be assumed to be equal to N_{22}. To find P_{22}, Eq. 10 is applied. The four receptances required to populate G_{22} are now known and Eq. 8 can be used to obtain R_{3b3b}. Given R_{3b3b}, free-free models for arbitrary tool-holder combinations can be developed and coupled to the spindle-machine receptances to predict the tool point FRF, H_{11}, required for milling process simulation.

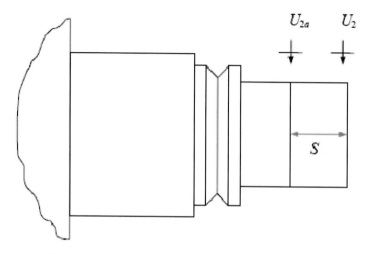

Figure 6. Locations for direct and cross artifact-spindle-machine assembly measurements used to calculate N_{22}.

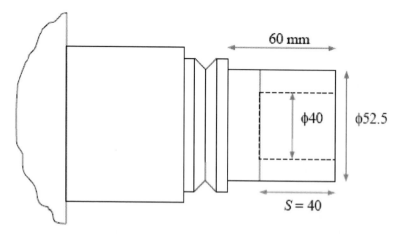

Figure 7. Artifact dimensions for Cincinnati FTV5-2500 measurements.

$$P_{22} = \frac{\Theta_2}{M_2} = \frac{F_2}{X_2}\frac{X_2}{M_2}\frac{\Theta_2}{F_2} = \frac{1}{H_{22}}L_{22}N_{22} = \frac{N_{22}^2}{H_{22}} \qquad (10)$$

To demonstrate the RCSA approach, artifact measurements (HSK-63A holder-spindle connection) were performed on a Cincinnati FTV5-2500 CNC milling machine to isolate the spindle-machine receptances. The artifact dimensions are provided in Figure 7 ($S = 40$ mm). The spindle-machine receptances were determined using Eq. 8 as described previously. Predictions and measurements were then completed for three tool-holder combinations. In each case, the holder was a Haimer A63.140 shrink fit chuck for 12.7 mm diameter endmills. Two different Data Flute three-flute solid carbide endmill models were tested at three overhang lengths (from the holder face to the endmill free end) to compare measurements and predictions. The 12.7 mm shank diameter HVM-30500 endmills had an overall length of 76.2 mm, a flute length of 15.9 mm, a relieved neck diameter of 12.0 mm, and length below shank of 34.9 mm. The 12.7 mm shank diameter HVM-M-30500 endmill had an overall length of 101.6 mm, a flute length of 15.9 mm, a relieved neck diameter of 12.0 mm, and length below shank of 54.0 mm. Two HVM-30500 endmills were tested with overhang lengths of 38.1 mm and 50.8 mm. One HVM-M-30500 endmill was tested with an overhang length of 63.5 mm.

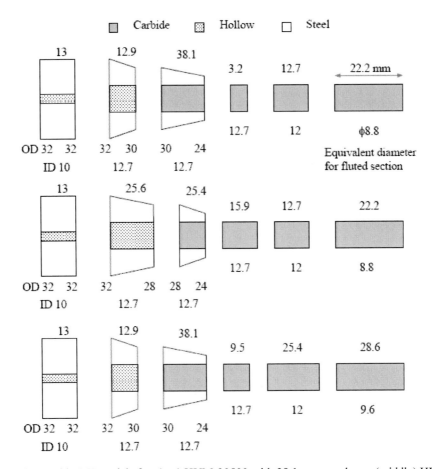

Figure 8. Subassembly I-II models for: (top) HVM-30500 with 38.1 mm overhang; (middle) HVM-30500 with 50.8 mm overhang; (bottom) HVM-M-30500 with 63.5 mm overhang.

The geometries for the free-free subassembly I-II models are provided in Figure 8. The equivalent diameter for the fluted section was calculated using the tool mass, an assumed carbide density of 14500 kg/m^3, the shank diameter and length, the neck diameter and length, and the length of the ground flutes. For the steel sections of the Timoshenko beam models, the elastic modulus was 200 GPa, the density was 7800 kg/m^3, and Poisson's ratio was 0.29. For the carbide sections, the elastic modulus was 550 GPa, the density was 14500 kg/m^3, and Poisson's ratio was 0.22. The unitless structural (solid) damping factor was 0.0015 for all sections.

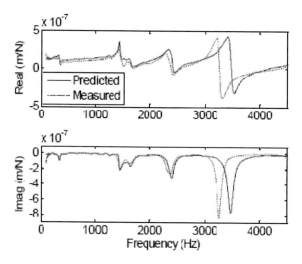

Figure 9. Comparison of measurement and prediction for HVM-30500 endmill with overhang of 38.1 mm.

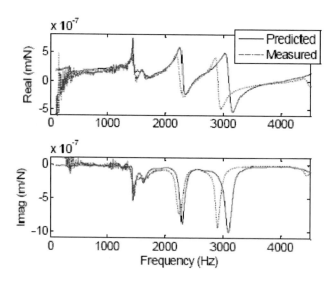

Figure 10. Comparison of measurement and prediction for HVM-30500 endmill with overhang of 50.8 mm.

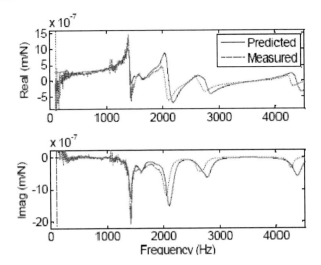

Figure 11. Comparison of measurement and prediction for HVM-M-30500 endmill with overhang of 63.5 mm.

Comparisons between the measured and predicted tool point FRF for the three tool-holder-spindle-machine assemblies are provided in Figures 9-11. These results are for measurements in the machine x direction. Similar agreement was obtained for the y direction. As noted previously, these tool point FRF predictions can be used as input to milling dynamics predictive algorithms (for stability and SLE) in order to select preferred operating parameters at the process planning stage.

4. PARAMETER SELECTION USING THE MILLING SUPER DIAGRAM

In its initial implementation, the super diagram combined stability and SLE information without considering the effects of tool wear [29]. To construct the diagram, the user selected the radial depth of cut, feed per tooth, SLE limit, and spindle speed-axial depth domain. Other inputs included the tool point FRF and force model. The spindle speed-axial depth domain was discretized into a grid of test points and a penalty was applied to each of the points as follows. Points that were predicted to be stable and within SLE tolerance levels were not penalized; their value was set to zero. Points that were stable, but outside the SLE tolerance were penalized by one (value = -1). Unstable points were penalized by two (value = -2). The point values for the selected grid were then used to construct a contour plot that served as the super diagram. A gray-scale scheme was used to separate the different zones. The feasible zone (point values = 0) was white, the SLE-limited zones (-1) were gray, and the unstable zone (-2) was black.

In order to incorporate the effects of tool wear, the cutting force model provided in Eq. 11 is experimentally evaluated as a function of the wear status of the tool. This enables a diagram to be constructed for any level of tool wear depending on the volume of material removed.

$$F_t = K_t bh + K_{te} b$$
$$F_n = K_n bh + K_{ne} b \qquad (11)$$

In Eq. 11, F_t is the tangential force component, K_t is the tangential cutting force coefficient, b is the axial depth of cut, h is the instantaneous chip thickness (which depends on the feed per tooth, f_t), K_{te} is the tangential edge (plowing) coefficient, F_n is the normal force component, K_n is the normal cutting force coefficient, and K_{ne} is the normal edge coefficient [6-7]. It has been previously shown that the force model coefficients tend to increase as wear progresses, e.g., [30]. By correlating the change in these coefficients with wear status, the diagram can be tailored to the behavior of a new tool or one at or near its end of life. The limit between stable/unstable behavior, as well as between acceptable/unacceptable SLE values, was previously indicated using a binary format. To incorporate uncertainty into the super diagram, a user-dependent safety margin is applied to modify the feasible (white) zone. The user selects how close in spindle speed and axial depth he/she is willing to operate relative to the predicted limits. Points within the feasible zone which violate this margin are penalized and a new "safe" feasible zone is identified. A second gray level is now incorporated. Dark gray indicates the stable points where the SLE limit is exceeded, while light gray represents the previously feasible points which violate the safety margin.

Table 1. Super diagram numerical case study parameters

Parameter	Value	Units
Stiffness	5×10^7	N/m
Damping ratio	0.05	-
Natural frequency	300	Hz
Tool diameter	19.05	mm
Helix angle	0	degrees
Number of teeth	1	
Tangential cutting coefficient	2×10^9	N/m^2
Normal cutting coefficient	0.667×10^9	N/m^2
Feed per tooth	0.06	mm/tooth
Radial depth of cut	19.05	mm

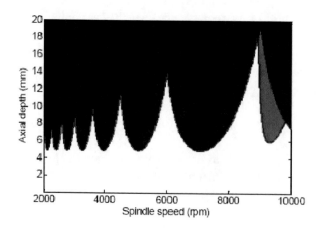

Figure 12. Super diagram with 25 μm SLE limit; the effects of tool wear are not included.

To demonstrate the super diagram including tool wear and uncertainty, a numerical case study is presented. The parameters are provided in Table 1. The original diagram for the selected system, which does not consider tool wear, is shown in Figure 12, where the surface location error limit is 25 μm.

Tool life is traditionally specified by the time required to reach a pre-selected wear level, often quantified using flank wear width (FWW). The FWW tends to increase with volume removed and is cutting speed (spindle speed) dependent. The assumed relationship between FWW and volume removed for the numerical example is shown in Figure 13, where the tool life is defined as the time to reach a FWW of 0.3 mm.

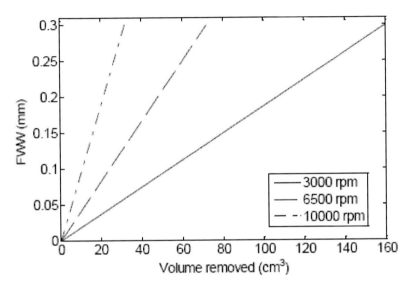

Figure 13. Variation in FWW with volume removed at different spindle speeds.

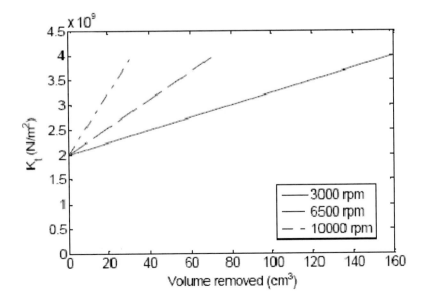

Figure 14. Variation in K_t with volume removed at different spindle speeds.

As noted, the cutting forces tend to grow with FWW. The assumed relationships between the cutting coefficients, K_t and K_n, identified in Eq. 11 and the volume removed are provided in Figures 14 and 15. The assumed linear relationship between these coefficients, spindle speed, Ω, and volume removed, V, is provided in Eq. 12, where the intercepts, $c_{0,t}$ and $c_{0,n}$, are the new tool coefficients (Table 1), and $c_{1,t}$ and $c_{1,n}$ are the speed-dependent rates of increase with V.

$$K_t(\Omega, V) = c_{0,t} + c_{1,t} V$$
$$K_n(\Omega, V) = c_{0,n} + c_{1,n} V \tag{12}$$

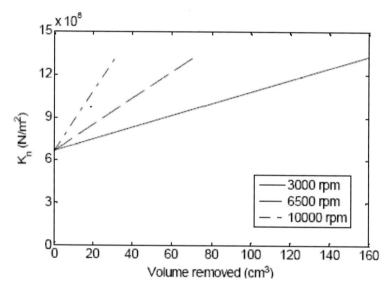

Figure 15. Variation in K_n with volume removed at different spindle speeds.

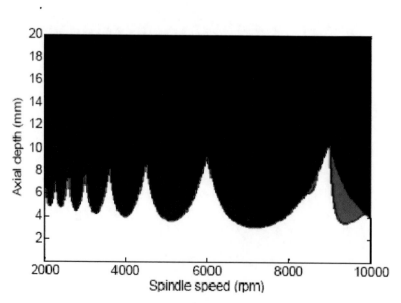

Figure 16. Super diagram with tool wear effects included ($V = 32$ cm^3).

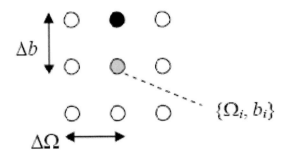

Figure 17. The safety margin is identified by testing the feasibility of the eight grid points surrounding $\{\Omega_i, b_i\}$. In this case, the (middle) test point is penalized (-1) because the (black) point above it is unstable.

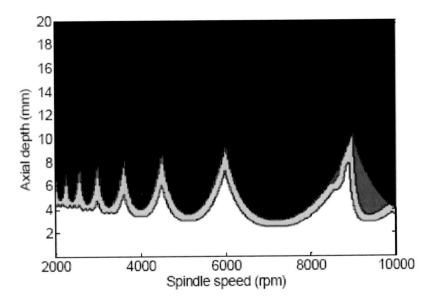

Figure 18. Super diagram including both tool wear effects and the user-defined safety margin ($V = 32$ cm^3, $\Delta\Omega = 100$ rpm, and $\Delta b = 0.5$ mm).

These slopes were assumed to increase linearly with Ω between 2000 rpm and 10000 rpm such that the coefficients doubled at 10000 rpm for $V = 32$ cm^3 (where FWW = 0.3 mm; see Figure 13) with no change at 2000 rpm for the same V. As an example, for $\Omega = 3000$ rpm with $V = 32$ cm^3, $c_{1,t} = \dfrac{c_{0,t}}{V}\dfrac{3000-2000}{10000-2000} = 7.8125\times10^6$ N/m^2/cm^3, by linear interpolation and $K_t = 2.0\times10^9 + 7.8125\times10^6 V = 2.25\times10^9$ N/m^2. Given this relationship between cutting coefficients, V, and Ω, the super diagram can then be modified to incorporate tool wear (the edge coefficients are not included in this example without loss of generality). As before, the $\{\Omega, b\}$ domain is represented by a grid of points and the stability and SLE is determined for each point. However, in this case, the volume to be removed must first be selected by the user. Then, the coefficients can be calculated for each spindle speed and, subsequently, the stability limit can be determined. Also, SLE is calculated at each axial depth grid point for the given spindle speed. The new diagram for $V = 32$ cm^3 is provided in Figure 16. Because the cutting coefficients grow with Ω, the stability limit decreases and the SLE infeasible zone grows while moving from left to right in the diagram.

The super diagram may also be modified to incorporate the user's beliefs regarding uncertainty in the actual location of the deterministic boundaries. To carry out this task, the user defines safety limits for spindle speed, $\Delta\Omega$, and axial depth of cut, Δb. These values give the distances from the boundaries that represent his/her 95% confidence level for actual feasible performance. For each feasible point in the $\{\Omega, b\}$ domain defined by the white zone, the penalty value of the surrounding eight points at distances $\Delta\Omega$ and Δb from the test point are queried; see Figure 17. If any of these points are infeasible (with a penalty of -1 or -2), then the test point is penalized and also identified as infeasible.

A new gray-scale is then implemented where the point values are: feasible (0, white), safety margin (-1, light gray), SLE limit (-2, dark gray), and unstable (-3, black). Therefore, the (white) feasible zone is reduced after the application of the user-specified safety margins. Figure 18 shows a super diagram from Figure 16 with safety margins of $\Delta\Omega$ = 100 rpm and Δb = 0.5 mm applied.

Given the feasible domain identified in Figure 18, a profit maximization strategy can be implemented to identify the optimized $\{\Omega, b\}$ combination. If a fixed revenue per part is assumed, then profit is maximized by minimizing the cost per part. The machining cost per part depends on both the operating parameters and the selected part path. For this example, a multi-layer zig-zag path strategy is selected to produce a 4 cm x 4 cm x 2 cm deep pocket (V = 32 cm^3); see Figure 19. The commanded stepover is 25% of the tool diameter (4.76 mm). However, because the first pass in each layer must be completed under slotting conditions, i.e., the radial depth equal to the tool diameter, the super diagram provided in Figure 18 (for slotting) is still appropriate. This ensures that the first pass is stable, but provides conservative conditions for the remaining passes.

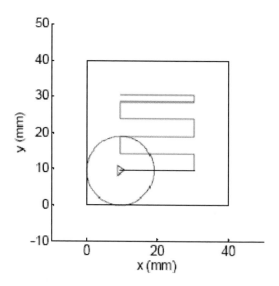

Figure 19. Representation of the part path for a single layer of the commanded pocket.

The pocket machining cost is calculated here as the product of the machining time and the cost rate ($2 per minute is assumed). The machining time depends on the cutting parameters (axial and radial depths, spindle speed, and feed per tooth), part path, and acceleration/deceleration values for the machining center. The zig-zag path is defined as follows:

1. The rotating tool begins at a position 10 mm above (z direction) the starting point of the pocket.
2. The tool approaches the part at the machining feedrate (the product of the feed per tooth, number of teeth, and spindle speed) and plunges into the material to the desired axial depth of cut.
3. The first machining pass is performed in the x direction at the edge of the pocket. This is a slotting cut for each layer.
4. At the end of the x direction pass, the tool is moved in the y direction to the desired radial depth and is then moved in the x direction for the next cutting pass. As shown in Figure 19, both up and down-milling is applied in each layer, i.e., there are cutting passes in both the +x and –x directions.
5. Step 4 is repeated until the full width of the pocket in the y direction is reached. The radial depth for the final x direction pass is less than the previous four passes (25% radial immersion) for the selected tool diameter and pocket dimensions.
6. When the layer is complete, a finishing pass around the pocket periphery is completed.
7. After the finishing pass, the tool is retracted to a rapid plane 10 mm above the completed layer and then moved to the pocket origin to begin the next layer; both moves are completed at the rapid traverse feedrate (300 mm/s). A z direction plunge cut at the machining feedrate is used to reach the desired axial depth for the next layer. This process is repeated until the pocket is complete; the tool is finally retracted to the original rapid plane.

$$t_a = \frac{f}{a} \tag{13}$$

To calculate the machining time for any prescribed motion, there are two primary steps. First, the acceleration time, t_a, required to reach the commanded feedrate (or decelerate from the commanded feedrate to zero velocity) is calculated using Eq. 13, where f is the commanded feedrate and a is the acceleration/deceleration (assumed equal). Given t_a, the associated distance, d_a, required to reach the commanded velocity is:

$$d_a = \frac{1}{2}at_a^2, \tag{14}$$

where the initial velocity is zero. By comparing d_a to one half of the commanded linear distance (to account for both acceleration at the motion start and deceleration at the motion end), it is determined if the tool reaches the commanded feedrate. If d_a is less than or equal to half the commanded distance, the commanded feedrate is reached. Otherwise, the velocity is constantly increasing for the first half of the move and constantly decreasing for the second half. The second step is to sum the acceleration/deceleration time ($2t_a$) with the constant velocity time, t_v, (if applicable):

$$t_v = \frac{d - 2d_a}{f}, \tag{15}$$

where d is the total move distance (for a y direction stepover, x direction pocket cut, z direction plunge cut, rapid traverse motion, or x/y direction finishing pass). The total machining time, t_t, for a given move is then either:

$$t_t = 2t_a + t_v, \tag{16}$$

if the commanded feedrate is reached or:

$$t_t = 2\sqrt{\frac{d}{a}}, \tag{17}$$

if the commanded feedrate is not reached.

A related consideration in the time calculation is the number of stepovers required in the y direction for each layer and the number of layers (and corresponding plunges in the z direction). The number of y direction steps, N_y, is calculated using the y direction dimension, $L_y = 40$ mm, the diameter of the tool, $D = 19.05$ mm, and the stepover (radial depth of cut), $s = 4.7625$ mm; see Eq. 18. Because the number of steps from Eq. 18 is not an integer for the selected pocket, the decimal portion is multiplied by s to obtain the final stepover distance ($0.399 \cdot 4.7625 = 1.9$ mm). A similar calculation is performed in the z direction; the result depends on the selected axial depth of cut.

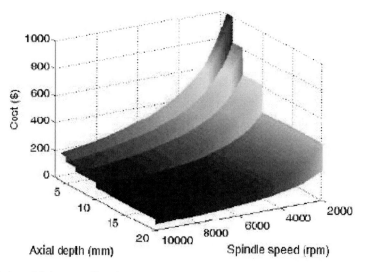

Figure 20. Pocket machining cost function.

The x direction moves traverse the full pocket dimension, $L_x = 40$ mm, a total of six times. Using this information, the machining and rapid feedrates, and the machine's acceleration/deceleration, the pocket machining time is determined as described previously.

$$N_y = \frac{L_y - D}{s} = \frac{40 - 19.05}{4.7625} = 4.399 \tag{18}$$

The layered nature of the part path yields an interesting effect on the pocket machining time and, therefore, the cost function; see Figure 20. Steps in the cost per part value are observed at axial depths which are integer fraction of the pocket depth: {20, 10, 6.667, 5, 4, ...} mm. For example, consider a limiting axial depth of 8 mm (to satisfy both the stability and SLE constraints). To complete the 20 mm deep pocket, three layers are required (two at 8 mm and one at 4 mm). If the limiting axial depth was increased to 9 mm and the same spindle speed was used, the cost would not be changed because three layers at the same feedrate would still be required. However, if the axial depth is increased to 10 mm, the number of layers decreases to two and the cost drops accordingly. The effect of spindle speed at any axial depth is seen as the decrease in cost with increasing spindle speed (and feedrate for a fixed feed per tooth value).

By calculating the machining cost at each point in the super diagram's feasible domain (Figure 18), the optimum machining conditions can be selected. For the new tool, the super diagram from Figure 12 is augmented with the safety margin ($\Delta\Omega = 100$ rpm and $\Delta b = 0.5$ mm) and the optimum conditions are identified by a pair of triangles which indicate the parameters with the lowest machining cost. The spindle speed is 8876 rpm and range of axial depths is 10 mm to 11.77 mm (this range occurs due to the stepped nature of the cost function). The associated costs are \$87.46 for 0.981 m/s^2 acceleration and \$85.66 for 9.81 m/s^2 acceleration. For the worn tool ($V = 32$ cm^3), the optimum spindle speed is 8903 rpm and the range of axial depths is 6.69 mm to 7.83 mm. The corresponding costs are \$129.10 (0.981 m/s^2) and \$125.98 (9.81 m/s^2). See Figure 22.

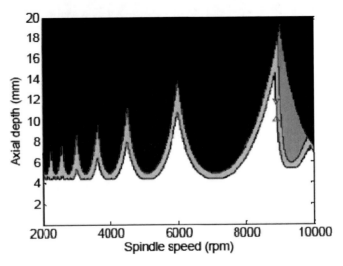

Figure 21. Optimum operating parameters for new tool with 25 µm SLE limit and user-defined safety margin ($\Delta\Omega = 100$ rpm and $\Delta b = 0.5$ mm).

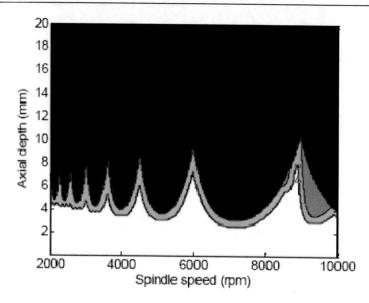

Figure 22. Optimum operating parameters for worn tool ($V = 32$ cm^3) with 25 μm SLE limit and user-defined safety margin ($\Delta\Omega = 100$ rpm and $\Delta b = 0.5$ mm).

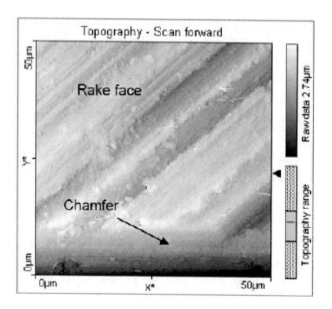

Figure 23. Typical AFM measurement of carbide insert rake face.

5. EXPERIMENTAL CASE STUDY

To evaluate the variation in stability behavior with tool wear, tool wear data for a 19 mm diameter inserted endmill (one square uncoated Kennametal 107888126 C9 JC carbide insert; zero rake and helix angles, 15 deg relief angle, 9.53 mm square x 3.18 mm) was collected. The part material was 1018 steel. An atomic force microscope (AFM) was used to measure the surface topography of the new carbide inserts. Figure 23 shows an example 50 μm x 50 μm measurement (256 line scans, no digital filtering) of the rake face. It is seen that there is a small chamfer with a 167 deg angle at the cutting edge. The roughness average for the rake face is 310 nm.

The first test was completed at a spindle speed, Ω, of 2500 rpm with a 3 mm axial depth of cut and 4.7 mm radial depth of cut (25% radial immersion). The feed per tooth value was 0.06 mm/tooth. The four coefficients in Eq. 11 were evaluated by performing a linear regression to the mean x (feed) and y direction forces obtained over a range of feed per tooth values: f_t = {0.03, 0.04, 0.05, 0.06, and 0.07} mm/tooth [6-7]. The cutting forces were monitored using a table-mounted force dynamometer (Kistler 9257B). In addition to calculating the cutting force coefficients intermittently while wearing the tool, the insert wear profile was also recorded at these intervals. To avoid removing the insert/tool from the spindle, a handheld microscope (60x magnification) was fixtured inside the machine enclosure and was used to measure the rake and flank surfaces. The calibrated digital images were used to identify the flank wear width (FWW). No crater wear was observed.

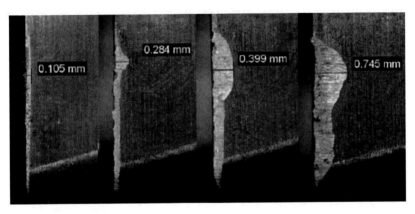

Figure 24. Images of insert relief face at 60x magnification (from left to right, V = {50, 125, 200, and 275} cm^3).

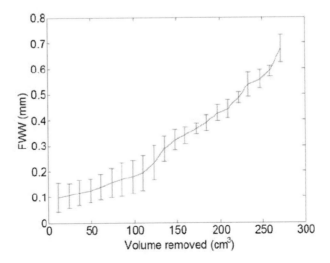

Figure 25. Variation in FWW with volume removed (Ω = 2500 rpm).

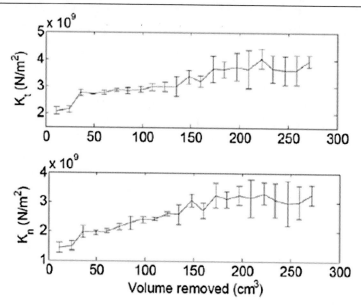

Figure 26. Variation in K_t and K_n with volume removed (Ω = 2500 rpm).

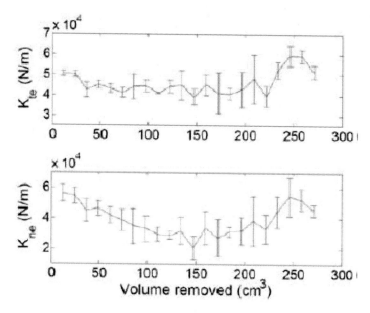

Figure 27. Variation in K_{te} and K_{ne} with volume removed (Ω = 2500 rpm).

Microscope images of the relief face for selected volumes of material removed, V, are shown in Figure 24. The sequence of tests was repeated three times to evaluate the repeatability from one insert to the next. Figure 25 shows the increase in maximum FWW with volume removed; one standard deviation (1σ) error bars are also included. The measurements were completed at increments of 12 cm^3. The force coefficients were also calculated using the linear regression approach at each interval. These results are shown in Figures 26 and 27.

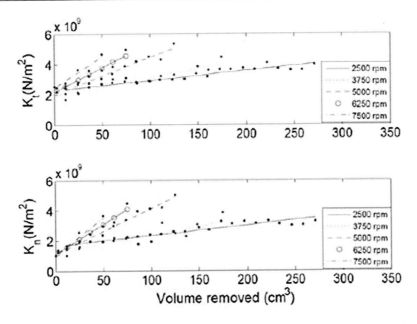

Figure 28. Variation in K_t and K_n with volume removed for $\Omega = \{2500, 3750, 5000, 6250, \text{and } 7500\}$ rpm.

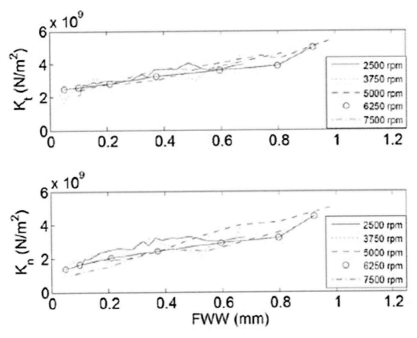

Figure 29. Variation in K_t and K_n with FWW for $\Omega = \{2500, 3750, 5000, 6250, \text{and } 7500\}$ rpm.

It is seen in Figure 26 that K_t and K_n increased with tool wear, while Figure 27 shows that the K_{te} and K_{ne} values exhibited no clear trend. Additionally, the distributions in K_t and K_n values increased with tool wear. For the selected tool/material pair, there was an approximately linear growth in K_t and K_n (R^2 values of 0.877 and 0.853 for K_t and K_n, respectively).

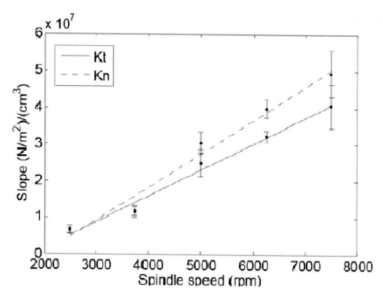

Figure 30. Variation in slope with spindle speed for the K_t and K_n versus volume removed lines from Figure 28.

Next, additional tests were carried out at spindle speeds of {3750, 5000, 6250, and 7500} rpm. The procedure and parameters (other than spindle speed) were the same as described for the 2500 rpm testing. The results are displayed in Figure 28; the linear least squares fits to the data are also shown. It is seen that the growth rates of K_t and K_n with volume removed (i.e., the slopes of the lines) increase with spindle speed. Interestingly, when plotted versus the corresponding FWW (measured with the microscope), the five different spindle speed results collapse onto a single line; see Figure 29. This suggests that if the FWW were monitored, it could provide an in-process approach to updating the force model coefficients based on the tool wear status. The K_{te} and K_{ne} values again did not exhibit any significant trend.

To describe the variation in K_t and K_n with volume removed as a function of spindle speed, the slopes of the individual lines in Figure 28 are plotted against spindle speed in Figure 30. As seen from the figure, the slopes increase linearly with spindle speed. The error bars in the figure were obtained from Monte Carlo simulation, where random values of K_t and K_n (within the 1σ range) were selected for each volume and a line was fit to this combination. The slope was calculated for each line and the mean and standard deviation for each spindle speed was used to construct Figure 30.

Using Figures 28 and 30, the increase in force coefficients K_t and K_n with spindle speed and volume removed can be expressed using the linear relationships given by Eq. 12. In this equation, $c_{0,t}$ and $c_{0,n}$, are the coefficient values for a new tool ($V = 0$ in Figure 26). The speed-dependent rates of increase, $c_{1,t}$ and $c_{1,n}$, are calculated using the slopes and intercepts of the lines in Figure 30. The slopes are 7.1×10^3 (N/m^2/cm^3)/rpm and 9.1×10^3 (N/m^2/cm^3)/rpm and the intercepts are -1.3×10^7 N/m^2/cm^3 and -1.8×10^7 N/m^2/cm^3 for the K_t and K_n data, respectively. The negative intercept values are attributed to the linear fit with inherent experimental uncertainty. The terms $c_{1,t}$ and $c_{1,n}$ are defined at different speeds by multiplying the slope by the corresponding spindle speed and adding the intercept as shown in Eq. 19. For the given tool-material combination, the increase in force coefficients at any spindle speed-volume removed combination within the testing range can be calculated using this equation. The coefficients can then be used to develop a super diagram that accounts for tool wear effects.

$$K_t(\Omega,V) = c_{0,t} + (7.1 \times 10^3 \Omega - 1.3 \times 10^7)V$$
$$K_n(\Omega,V) = c_{0,n} + (9.1 \times 10^3 \Omega - 1.8 \times 10^7)V \qquad (19)$$

The previous tests were performed at f_t = 0.06 mm/tooth. This enables the coefficients to be evaluated at that value. However, changing the feed per tooth can affect the wear rate and SLE values. Therefore, a similar set of experiments was completed at f_t = {0.03, 0.045, 0.075, and 0.09} mm/tooth. The tests were completed at 5000 rpm with all other parameters held constant. Figure 31 shows the variation in cutting force coefficients with volume removed for these feed per tooth values.

The wear rate is higher and the volume of material that can be removed is lower for the smaller feed per tooth values. This wear rate trend suggests that strain hardening may be in effect. The thinner chips with increased hardness can cause accelerated wear. The reduced amount of material that can be removed could also be attributed to the increase in cutting time and the number of passes through the material required to remove the same volume for smaller feed per tooth values.

Finally, the variation in wear rate behavior with axial depth of cut was evaluated. The axial depths were {3, 4.5, and 6} mm, the spindle speed was 5000 rpm, the feed per tooth was 0.06 mm/tooth and the radial depth remained at 4.7 mm. Figure 32 shows the results. Note that the K_t and K_n values are plotted against volume normalized by the axial depth of cut, $V_n = V/b$. This normalization was necessary because the independent variable, V, is a function of the dependent variable, b. As seen in the figure, the three test sets collapse onto a single line for the usable tool life when plotted versus the normalized volume. It has also been suggested that variation in FWW is not observed at different radial depths of cut and differing number of teeth (assuming no runout) [22].

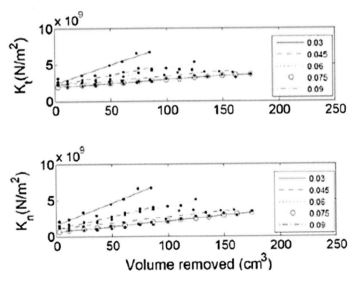

Figure 31. Variation in K_t and K_n with volume removed for f_t = {0.03, 0.045, 0.06, 0.075 and 0.09} mm/tooth.

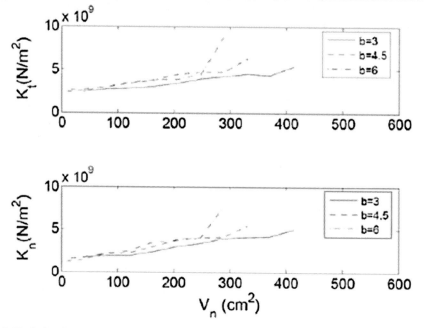

Figure 32. Variation in K_t and K_n with normalized volume removed for $b = \{3, 4.5, \text{ and } 6\}$ mm.

By normalizing the volume removed by the axial and radial depths of cut and number of teeth, the required number of tests can be substantially reduced. For a given tool-part material combination, testing can therefore be completed only at a selected axial depth of cut, radial depth of cut, and number of teeth. The results can then be extended to other combinations by plotting the values of force coefficients against the normalized volume removed.

The tool wear experimental results showed a linear increase in cutting force coefficients K_t and K_n with volume removed due to progressive flank wear. The rate of increase of force coefficients increased linearly with speed. This increase in force coefficients causes the limiting axial depth of cut for stable milling to decrease. To explore this wear effect, the tool point frequency response function was measured by impact testing and the stability limit was calculated using force coefficients based on a new and worn insert. An insert was worn by removing 275 cm^3 at $\Omega = 2500$ rpm. The force coefficients were determined for both a new insert and the worn insert using a linear regression of the average x and y direction forces at varying feed per tooth values as described previously. These force coefficients are provided in Table 2. For stability testing at 5100 rpm, the equivalent volume removed which would yield the K_t and K_n values for the worn insert was calculated to be 121 cm^3 using Eq. 19. The stability limit for the new insert was calculated using the new insert values ($V = 0$). For the worn insert, the K_t and K_n values at each spindle speed were calculated using Eq. 20 and the stability limit was generated as described in the numerical study. However, as shown by the error bars in Figure 30, there is uncertainty in the K_t and K_n values for both the new and worn inserts. A Monte Carlo simulation was completed where random K_t and K_n values were selected from the experimental distributions and a new stability limit was calculated for each set. See Figure 33, where the radial depth of cut is equal to the tool diameter (slotting). The band of stability limits indicates the uncertainty. This information could be used, for example, to aid a user in selecting his/her safety limits for the super diagram. The mean limiting depth of cut at 5100 rpm is 2.15 mm for the new insert and 0.85 mm for the worn insert.

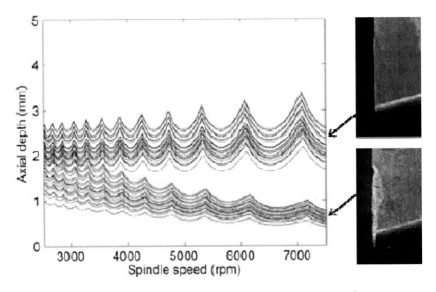

Figure 33. Stability lobe diagrams for new ($V = 0$) and worn inserts ($V = 121$ cm^3).

Table 2. Force coefficient values for new and worn inserts

	K_t (N/m^2)	K_n (N/m^2)	K_{te} (N/m)	K_{ne} (N/m)
New	1.90×10^9	0.78×10^9	45500	46650
Worn	4.98×10^9	4.51×10^9	45500	25500

$$K_t(\Omega,V) = 2.2 \times 10^9 + (7.1 \times 10^3 \Omega - 1.3 \times 10^7)121$$
$$K_n(\Omega,V) = 1.2 \times 10^9 + (9.1 \times 10^3 \Omega - 1.8 \times 10^7)121 \quad (20)$$

Cutting tests were completed at $b = \{0.8, 1.6, 2.2,$ and $3\}$ mm with the new and worn inserts. A once-per-revolution force sampling strategy for the x (F_x) and y (F_y) directions was used to identify chatter. The once-per-revolution samples were obtained by sampling the force data at the commanded spindle rotating frequency. For stable cutting conditions, the once-per-revolution samples (due to forced vibration only) are synchronous with spindle rotation and produce a small cluster of points [47] in the F_x vs. F_y plot. Unstable behavior, on the other hand, produces a more distributed set of points due to its asynchronous nature. A statistical variance ratio, R, is used as an indicator of chatter [43]; see Eq. 21, where $\sigma^2_{opr,x}$ and $\sigma^2_{opr,y}$ are the variances in the once-per-revolution sampled forces in the x and y directions and σ^2_x and σ^2_y are the variances in the x and y direction forces. Figure 34 shows once-per-revolution samples for tests at $b = 1.6$ mm and $\Omega = 5100$ rpm for the new and worn inserts.

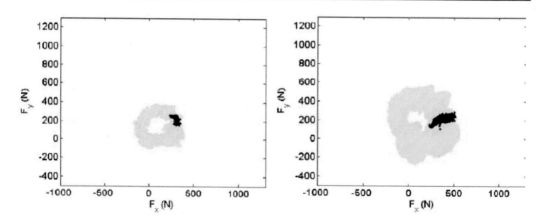

Figure 34. Once-per-revolution samples for 1.6 mm axial depth of cut and 5100 rpm spindle speed (left: new insert, right: worn insert).

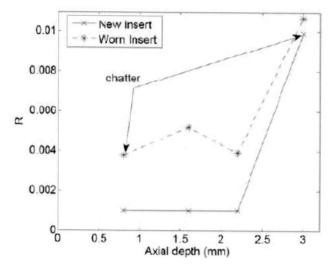

Figure 35. Statistical variance ratio, R, for new and worn inserts at different axial depths of cut.

$$R = \frac{\sigma_{opr,x}^2 + \sigma_{opr,y}^2}{\sigma_x^2 + \sigma_y^2} \qquad (21)$$

As seen in Figure 34, the distribution of once-per-revolution samples increases for the unstable cut using the worn insert. Similar results were obtained $b = 0.8$ mm (which is within the stability limit distribution for the worn tool in Figure 33), 2.2 mm, and 3 mm. Figure 34 shows the R values for different axial depths of cut, where the R value is larger for all cuts with the worn insert and increases substantially for both the new and worn insert at $b = 3$ mm (violent chatter occurred in both cases for this depth). Sample surface profiles for $\Omega = 5100$ rpm with $b = 1.6$ mm for the new and worn inserts corroborate the R results. Figures 36 and 37 display the topography of the machined surface obtained using a scanning white light interferometer with a 10x magnification and 2.5 mm by 1 mm field of view. The unstable condition for the worn insert produces the expected rough surface finish. The average surface roughness for Figure 36 is 781.8 nm; this surface does not show any distinct chatter marks. The average surface roughness for the unstable result (worn tool) in Figure 37 is 4018.5 nm (5.1 times increase) and the surface exhibits chatter marks. These results indicate that the stability limit did decrease with increased tool wear.

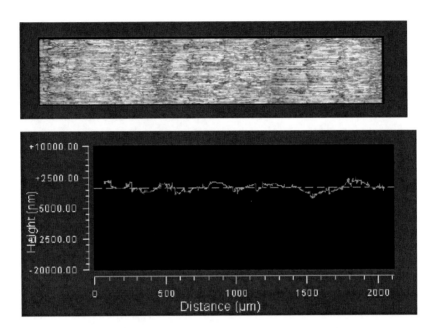

Figure 36. Surface profile for new insert (stable cutting) for 5100 rpm with $b = 1.6$ mm.

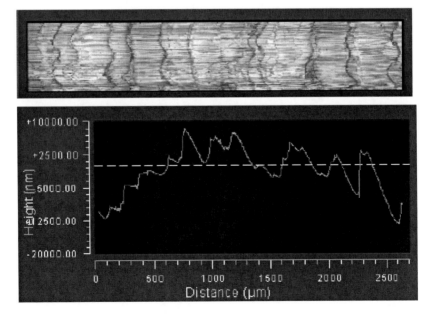

Figure 37. Surface profile for worn insert (unstable) for 5100 rpm with $b = 1.6$ mm.

6. CONCLUSION

This chapter described the application of structural and process dynamics models to the pre-process selection of optimum milling operating parameters for increased process performance. The Receptance Coupling Substructure Analysis (RCSA) method was used to predict the tool point frequency response function. Given the tool point response and an appropriate force model that relates the cutting force to chip area for the cutting tool-part material pair, the stability and surface location error (due to forced vibrations) for the selected milling operation were predicted. This information was presented graphically in the form of the milling "super diagram", which also: 1) included the effect of tool wear through a dependence of the force model coefficients on the volume of material removed and spindle speed; and 2) incorporated uncertainty in the form of user-defined safety margins. Given this information at the process planning stage, the programmer is then able to select optimum operating parameters that increase the likelihood of first part correct production and reduce the probability of damage to the tool, spindle, and/or part due to excessive forces and deflections.

ACKNOWLEDGMENTS

The authors gratefully acknowledge financial support from the National Science Foundation (CMMI-0926667 and CMMI-0928211) and Kennametal (donation of carbide inserts). The research was also sponsored by the U.S. Army Benet Laboratories and was accomplished under Cooperative Agreement Number W15QKN-06-2-0100. The views and conclusions contained in this document are those of the authors and should not be interpreted as representing the official policies, either expressed or implied, of U.S. Army Benet Laboratories or the U.S. Government. The U.S. Government is authorized to reproduce and distribute reprints for Government purposes notwithstanding any copyright notation heron.

REFERENCES

[1] Tobias, S.A. and Fishwick, W., 1958, The Chatter of Lathe Tools under Orthogonal Cutting Conditions, *Transactions of the ASME*, 80: 1079.

[2] Tlusty, J. and Polocek, M., 1963, The Stability of the Machine-Tool against Self-Excited Vibration in Machining, In: Proceedings of the International Research in Production Engineering Conference, Pittsburgh, PA, ASME, New York, NY, pp. 465.

[3] Merrit, H., 1965, Theory of Self-Excited Machine Tool Chatter, *Journal of Engineering for Industry*, Transactions of the ASME, 87/4: 447-454.

[4] Tlusty, J., Zaton, W., and Ismail, F., 1983, Stability Lobes in Milling, *Annals of the CIRP*, 32/1: 309-313.

[5] Altintas, Y. and Budak, E., 1995, Analytical Prediction of Stability Lobes in Milling, *Annals of the CIRP*, 44/1: 357-362.

[6] Altintas, Y., 2000, Manufacturing Automation, Cambridge University Press, Cambridge, UK.

[7] Schmitz, T. and Smith, K.S., 2009, Machining Dynamics: Frequency Response to Improved Productivity, Springer, New York, NY.

[8] Kline, W., DeVor, R., and Shareef, I., 1982, The Prediction of Surface Accuracy in End Milling, *Journal of Engineering for Industry*, 104: 272-278.

[9] Kline, W., DeVor, R., and Lindberg, J., 1982, The Prediction of Cutting Forces in End Milling with Application to Cornering Cuts, *International Journal of Machine Tool Design Research*, 22: 7-22.

[10] Tlusty, J., 1985, Effect of End Milling Deflections on Accuracy, In: R.I. King (Ed.), Handbook of High Speed Machining Technology, Chapman and Hall, New York, pp. 140-153.

[11] Sutherland, J. and DeVor, R., 1986, An Improved Method for Cutting Force and Surface Error Prediction in Flexible End Milling Systems, *Journal of Engineering for Industry*, 108: 269-279.

[12] Montgomery, D. and Altintas, Y., 1991, Mechanism of Cutting Force and Surface Generation in Dynamic Milling, *Journal of Engineering for Industry*, 113/2: 160-168.

[13] Smith, S. and Tlusty, J., 1991, An Overview of Modeling and Simulation of the Milling Process, *Journal of Engineering for Industry*, 113/2: 169-175.

[14] Altintas, Y., Montgomery, D., and Budak, E., 1992, Dynamic Peripheral Milling of Flexible Structures, *Journal of Engineering for Industry*, 114/2: 137-145.

[15] Tarng, Y., Liao, C., and Li, H., 1994, A Mechanistic Model for Prediction of the Dynamics of Cutting Forces in Helical End Milling, *International Journal of Modeling and Simulation*, 14/2: 92-97.

[16] Schmitz, T. and Ziegert, J., 1999, Examination of Surface Location Error Due to Phasing of Cutter Vibrations, *Precision Engineering*, 23/1: 51-62.

[17] Yun, W.-S., Ko, J., Cho, D.-W., and Ehmann, K., 2002, Development of a Virtual Machining System, Part 2: Prediction and Analysis of a Machined Surface Error, *International Journal of Machine Tools and Manufacture*, 42: 1607-1615.

[18] Mann, B., Bayly, P, Davies, M. and Halley, J., 2004, Limit Cycles, Bifurcations, and Accuracy of the Milling Process, *Journal of Sound and Vibration*, 277: 31–48.

[19] Schmitz, T., Couey, J., Marsh, E., Mauntler, N., and Hughes, D., 2007, Runout Effects in Milling: Surface finish, Surface Location Error, and Stability, *International Journal of Machine Tools and Manufacture*, 47: 841-851.

[20] Schmitz, T., Ziegert, J., Canning, J.S., and Zapata, R., 2008, Case Study: A Comparison of Error Sources in Milling, *Precision Engineering*, 32(2): 126-133.

[21] Schmitz, T. and Mann, B., 2006, Closed Form Solutions for Surface Location Error in Milling, *International Journal of Machine Tools and Manufacture*, 46: 1369-1377.

[22] Tlusty, J., 2000, Manufacturing Processes and Equipment. Prentice Hall, Upper Saddle River, NJ.

[23] Taylor, F.W., 1906, On the Art of Cutting Metals, *Transactions of the ASME*, 28: 31-248.

[24] Prickett, P. and Johns, C., 1999, An Overview of Approaches to End Milling Tool Monitoring, *International Journal of Machine Tools and Manufacture*, 39: 105–122.

[25] Rehorn, A., Jiang, J. and Orban, P., 2005, State-of-the-art Methods and Results in Tool Condition Monitoring: A Review, *International Journal of Advanced Manufacturing Technology*, 26: 693–710.

[26] Trent, E. and Wright, P., 2000, Metal Cutting 4th Ed., Butterworth Heinemann, Boston, MA.

[27] Dornfeld, D. and Lee, D.E., 2008, Precision Manufacturing, Springer, New York, NY.

[28] Slocum, A., 1992, Precision Machine Design, Prentice-Hall, Inc., Englewood Cliffs, NJ.

[29] Zapata, R., Karandikar J. and Schmitz T., 2009, A New "Super Diagram" for Describing Milling Dynamics, Transactions of NAMRI/SME, 36: 245-252.
[30] Cui, Y., Fussell, B., Jerard, R., and Esterling, D., 2009, Tool Wear Monitoring for Milling by Tracking Cutting Force Model Coefficients, Transactions of the NAMRI/SME, 37: 613-620.
[31] Ewins, D.J., 2000, Modal Testing: Theory, Practice and Application, 2nd Ed., Research Studies Press, Ltd., Philadelphia, PA.
[32] Bishop, R. and Johnson, D., 1960, The Mechanics of Vibration, Cambridge University Press, Cambridge.
[33] Jetmundsen, B., Bielawa, R., and Flannelly, W., 1988, Generalized Frequency Domain Substructure Synthesis, *Journal of the American Helicopter Society*, 33: 55-64.
[34] Schmitz, T. and Donaldson, R., 2000, Predicting High-Speed Machining Dynamics by Substructure Analysis, *Annals of the CIRP*, 49/1: 303-308.
[35] Schmitz, T., Davies, M., and Kennedy, M., 2001, Tool Point Frequency Response Prediction for High-Speed Machining by RCSA, *Journal of Manufacturing Science and Engineering*, 123: 700-707.
[36] Schmitz, T., Davies, M., Medicus, K., and Snyder, J., 2001, Improving High-Speed Machining Material Removal Rates by Rapid Dynamic Analysis, *Annals of the CIRP*, 50/1: 263-268.
[37] Burns, T., Schmitz, T., 2004, Receptance Coupling Study of Tool-Length Dependent Dynamic Absorber Effect, Proceedings of American Society of Mechanical Engineers *International Mechanical Engineering Congress and Exposition*, IMECE2004-60081, Anaheim, CA.
[38] Schmitz, T., and Duncan, G.S., 2005, Three-Component Receptance Coupling Substructure Analysis for Tool Point Dynamics Prediction, *Journal of Manufacturing Science and Engineering*, 127/4: 781-790.
[39] Duncan, G.S., Tummond, M., and Schmitz, T., 2005, An Investigation of the Dynamic Absorber Effect in High-Speed Machining, *International Journal of Machine Tools and Manufacture*, 45: 497-507.
[40] Cheng, C.-H., Schmitz, T., Arakere, N., and Duncan, G.S., 2005, An Approach for Micro End mill Frequency Response Predictions, Proceedings of American Society of Mechanical Engineers International Mechanical Engineering Congress and Exposition, IMECE2005-81215, Orlando, FL.
[41] Burns, T. and Schmitz, T., 2005, A Study of Linear Joint and Tool Models in Spindle-Holder-Tool Receptance Coupling, Proceedings of 2005 American Society of Mechanical Engineers International Design Engineering Technical Conferences and Computers and Information in Engineering Conference, DETC2005-85275, Long Beach, CA.
[42] Schmitz, T., Powell, K., Won, D., Duncan, G.S., Sawyer, W.G., Ziegert, J., 2007, Shrink Fit Tool Holder Connection Stiffness/Damping Modeling for Frequency Response Prediction in Milling, *International Journal of Machine Tools and Manufacture*, 47/9: 1368-1380.
[43] Cheng, C.-H., Duncan, G.S., and Schmitz, T., 2007, Rotating Tool Point Frequency Response Prediction using RCSA, *Machining Science and Technology*, 11/3: 433-446.
[44] Filiz, S., Cheng, C.-H, Powell, K., Schmitz, T., and Ozdoganlar, O., 2009, An Improved Tool-Holder Model for RCSA Tool-Point Frequency Response Prediction, Precision Engineering, 33: 26–36.
[45] Weaver, W., Jr., Timoshenko, S., and Young, D., 1990, Vibration Problems in Engineering, 5th Ed., John Wiley and Sons, New York, NY.

[46] Sattinger, S., 1980, A Method for Experimentally Determining Rotational Mobilities of Structures, Shock and Vibration Bulletin, 50: 17–27.

[47] Davies M., Dutterer B., Pratt J., Schaut A., and Bryan J., 1998, On the Dynamics of High-Speed Milling with Long, Slender Endmills, Annals of the, 47/1: 55-60.

Reviewed by Dr. John Snyder, TechSolve, Cincinnati, OH.

In: Machine Tools: Design, Reliability and Safety
Editor: Scott P. Anderson, pp. 153-173

ISBN: 978-1-61209-144-0
© 2011 Nova Science Publishers, Inc.

Chapter 5

OPTIMUM DESIGN OF A REDUNDANTLY ACTUATED PARALLEL MANIPULATOR BASED ON KINEMATICS AND DYNAMICS

Jun Wu[*], Jinsong Wang, Liping Wang and Tiemin Li

Department of Precision Instruments and Mechanology,
Tsinghua University, Beijing, China

ABSTRACT

Parallel manipulators have attracted much attention in both industry and academia because of their conceptual potentials in high motion dynamics and accuracy combined with high structural rigidity due to their closed kinematic loops. However, there is a gap between the expectation and practical application of parallel manipulators in the machine tool/robot sectors. One of the reasons is that their potentially desirable high dynamics can not be realized since the dynamic characteristics are not considered in the kinematic design phase. The dynamic characteristics can be considered in the model-based control after the prototype is built. However, once the prototype is fabricated, the improvement of dynamic characteristics is limited even if a model-based control is used. If the dynamic characteristics can also be involved in the process of the kinematic design before the prototype is built, the dynamic performance would be improved more. It is helpful to realize the high motion dynamics of parallel manipulators.

This chapter presents a new method for the optimum design of parallel manipulators by taking both the kinematic and dynamic characteristics into account. The optimum design of a 3-DOF redundant parallel manipulator with actuation redundancy is investigated to demonstrate the method. The dynamic model is derived and a dynamic manipulability index is proposed. Based on the results of kinematic optimal design, the manipulator is re-designed by considering the dynamic performance. The kinematic performance may be debased, but the dynamic performance is improved. By using the method proposed in this chapter, the designer can obtain the optimum result with respect to both kinematic performance indices and dynamic performance in dices. Since the dynamic performance is considered in the process of optimum design by using the

---

[*] E-mail: jhwu@mail.tsinghua.edu.cn.

method proposed in this chapter, it is expected to realize the high dynamics of parallel manipulators.

1. INTRODUCTION

Parallel manipulators have received great attention in the fields of high-speed machines and high-speed pick-and-place applications because of their conceptual potentials in high motion dynamics and accuracy combined with high structural rigidity due to their closed kinematic loops. In recent years, many studies have been dedicated to the improvement of parallel mechanisms, but a large part of it was devoted only to kinematics. In this field, important results have been obtained regarding analysis and synthesis. New results have also been obtained regarding dynamics and control. Many prototypes are developed to validate the results and the improvement of mechanism performance. However, there is still a gap from academia to industry. Only a few research prototypes can reach the robotics market [1,2]. One of the reasons is that their potentially desirable high dynamics can not be realized since the dynamic characteristics are not considered in the kinematic design. Although the dynamic characteristics can be considered in the model-based control after the prototype is built, the improvement of dynamic characteristics is limited. If the dynamic characteristics can also be involved in the kinematic design before the prototype is built, the dynamic performance will be improved. Thus, it is helpful to realize the high motion dynamics.

Optimal kinematic design is an important subject in designing a parallel manipulator. No matter how simple the parallel manipulator is, the optimum design is always challenging. Many efforts have been contributed to this issue. For example, Gao et al. [3] and Liu et al. [4-6] developed a solution space trying to solve the optimum design problem by means of atlases. Cervantes-Sanchez et al. [7] established a space made up of two normalized geometric parameters to show the characteristics of workspace and singularity. These approaches can be approximately classified into two kinds: one is based on the performance indices and the other is based on the performance atlases. The methods based on the performance indices are traditional ones, implemented in the commercial software MATLAB based on an established object function. If the design criterion number is increased, the procedure will be more complicated. The method based on the performance atlases can give designers a global and visual information on with what kind of link lengths the mechanism can have good or best performance. However, all these approaches only consider the kinematic performance. Once the prototype of a parallel manipulator is built, the dynamic improvement of the parallel manipulator is difficult to be realized. If the dynamic performance can also be involved in the process of the optimum design, the manipulator would have the desirable dynamics.

In order to study the dynamic characteristics of robot manipulators, some researchers defined some performance indices. Conventionally, dynamic manipulability ellipsoid (DME) [8,9] and generalized inertia ellipsoid (GIE) [10] were used as performance indices to evaluate the dynamic manipulability of a robot manipulator. Besides, some other measures for evaluating dynamic manipulability have been proposed. Graettinger and Krogh [11] introduced acceleration radius. Hashimoto [12] used the harmonic mean of the square singular values matrix to evaluate the dynamic manipulability. Li et al. [13] presented the smallest singular value of inertia matrix of a manipulator as the evaluation index when the manipulability in the hardest direction was considered. Zhao and Gao [14] defined a few performance indices to compare the dynamic performance of a 8PSS redundant parallel manipulator with its non-redundant counterpart. Mansouri and Ouali [15] presented a new homogenous manipulability measure of robot manipulators based on power concept. Nokleby et al [16] presented a methodology of using scaling factors to determine the force capabilities of non-redundantly and redundantly-actuated parallel manipulator. Chiacchio [17] presented a new definition of dynamic manipulability ellipsoid for redundant manipulators. However, most of these contributions concentrated on the analysis and evaluation of manipulator dynamic characteristics, and are not used for manipulator design to improve the dynamic performance. Although Asada [18] draw the GIE on a computer display and aided the design by visualizing the dynamic behaviour, it is not an applied design method.

The goal of this chapter is to propose a new method to take the dynamic characteristics of parallel manipulators into account in the process of optimum design. In reference [19], the kinematic optimization for the 3-DOF parallel manipulator with actuation redundancy is carried out. In the chapter, in addition to the kinematic optimization, the dynamic characteristics are also considered in the kinematic design. The dynamic model of the manipulator is derived and a performance index for evaluating dynamic manipulability is proposed. Based on the dynamic index, the manipulator is redesigned. The simulation results show that the dynamic manipulability of the manipulator is improved. By using the method proposed in the chapter, the designer can obtain the optimum result with respect to both kinematic performance indices and dynamic performance indices.

Figure1. 3-D model of the redundant parallel manipulator.

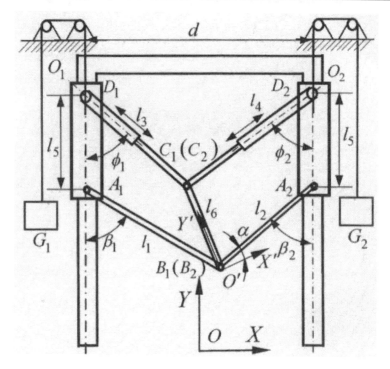

Figure 2. kinematic model of the redundant manipulator.

2. KINEMATIC ANALYSIS

2.1. Inverse Kinematics

The 3-D model and kinematic model of the redundantly actuated parallel manipulator are show in Figures 1 and 2, respectively. The parallel manipulator is composed of gantry frame, two active sliders, two constant length links, two extendible links and moving platform. In reference [19], it is incorporated into a hybrid machine tool. In the chapter, this redundant parallel manipulator is redesigned to obtain the desirable dynamic performance. Sliders are driven by two servomotors fixed on the columns via screw. Links $C_1 D_1$ and $C_2 D_2$, which are driven by two actuators, are extendible struts with one end fixed to sliders $A_1 D_1$ and $A_2 D_2$, and the other connected to the moving platform $B_1 C_1$. The whole construction enables movement of the moving platform in a plane and its rotation about the axis normal to the moving plane of the manipulator.

A base coordinate system $O-XY$ is shown in Figure 2, and a moving coordinate system $O'-X'Y'$ is fixed on the joint point B_1 and Y' axis along the vector from B_1 to B_2. The position vector of O' is defined as $r = [x \quad y]^T$ in the base coordinate system. The position vector of joint point $O_i (i=1,2)$, which is at the end of the column, is defined as $r_{Oi} = [0 \quad l_c]^T$ in the base coordinate system, and l_c is the height of the column.

Based on Fig 2, the following equations can be obtained

$$r_{Oi} + q_i e_2 + l_i s_i = r + R r'_{Bi} \quad i=1,2 \tag{1}$$

$$r_{Oi} + (q_i + l_5) e_2 + l_{i+2} s_{i+2} = r + R r'_{Ci} \tag{2}$$

where l_i is the length of link $A_i B_i$, l_{i+2} is the length of link $D_i C_i$, q_i is the distance from O_i to A_i, $e_2 = [0\ 1]^T$ is the unit vector along Y-axis, $s_i = [\sin\beta_i\ \cos\beta_i]^T$ and $s_{i+2} = [\sin\phi_i\ \cos\phi_i]^T$ are the unit vectors of links $A_i B_i$ and $D_i C_i$, and R is the rotation matrix from coordinate system $O' - X'Y'$ to $O - XY$ and $R = \begin{bmatrix} \cos\alpha & -\sin\alpha \\ \sin\alpha & \cos\alpha \end{bmatrix}$, α is the rotation angle of the moving platform, r'_{Bi} and r'_{Ci} are the position vectors of point $B_i (i=1,2)$ and $C_i (i=1,2)$ in $O' - X'Y'$.

Based on Eq. (1) and (2), the inverse solutions of the kinematics can be written as

$$q_i = e_2^T r_i \pm \sqrt{(e_2^T r_i)^2 - r_i^T r_i + l_i^2} \tag{3}$$

$$l_{i+2} = |r_{i+2} - (q_i + l_5) e_2| \tag{4}$$

where $r_i = r + R r'_{Bi} - r_{Oi}$ and $r_{i+2} = r + R r'_{Ci} - r_{Oi}$.

For the configuration shown in Fig 2, '$\pm$' in Eq. (3) should be only '$-$'.

2.2. Velocity Equation

Taking the time derivative of Eq. (1) and (2) leads to

$$\dot{q}_i e_2 + l_i \dot{\beta}_i E s_i = \dot{r} + \dot{\alpha} E R r'_{Bi}, \quad i=1,2 \tag{5}$$

$$\dot{q}_i e_2 + \dot{l}_{i+2} s_{i+2} + l_{i+2} \dot{\phi}_i E s_{i+2} = \dot{r} + \dot{\alpha} E R r'_{Ci} \tag{6}$$

where $E = \begin{bmatrix} 0 & 1 \\ -1 & 0 \end{bmatrix}$.

Multiplying both sides of Eq. (5) by s_i^T and e_1^T leads to

$$\dot{q}_i = J_i [\dot{x}\ \dot{y}\ \dot{\alpha}]^T \tag{7}$$

$$\dot{\beta}_i = \frac{e_1^T (\dot{r} + \dot{\alpha} E R r'_{Bi})}{l_i e_1^T E s_i} \tag{8}$$

where $J_i = \dfrac{s_i^T T_{Bi}}{s_i^T e_2}$, $T_{Bi} = [I_2 \quad ERr'_{Bi}]$, I_2 is a 2-dimension unit matrix, and $e_1 = [1 \quad 0]^T$.

Multiplying both sides of Eq. (6) by s_{i+2}^T and $s_{i+2}^T E^T$ leads to

$$\dot{l}_{i+2} = J_{i+2}[\dot{x} \quad \dot{y} \quad \dot{\alpha}]^T \tag{9}$$

$$\dot{\phi}_i = s_{i+2}^T E^T (\dot{r} + \dot{\alpha} ERr'_{Ci} - \dot{q}_i e_2)/l_{i+2} \tag{10}$$

where $J_{i+2} = s_{i+2}^T T_{Ci} - s_{i+2}^T e_2 J_i$ and $T_{Ci} = [I_2 \quad ERr'_{Ci}]$.

The Jacobian matrix J is defined as

$$\dot{q} = J\dot{p} \tag{11}$$

where $q = [q_1 \quad q_2 \quad l_3 \quad l_4]^T$ and $p = [x \quad y \quad \alpha]^T$.

Thus, the Jacobian matrix of the redundant parallel manipulator can be expressed as

$$J = [J_1^T \quad J_2^T \quad J_3^T \quad J_4^T]^T \tag{12}$$

3. KINEMATIC PERFORMANCE INDEX

It is well known that the velocity has its dimension. Since the output velocity of the manipulator includes the linear and angular velocity, the condition number of Jacobian matrix which is often used to evaluate the dexterity has no explicit physical meaning. To solve the problem, the angular velocity dimension and linear velocity dimension should be written as a uniform form. It is assumed that the moving platform rotates about joint point B_1. Eq. (11) can be rewritten as

$$\dot{q} = J'[\dot{y} \quad \dot{z} \quad l_6 \dot{\alpha}]^T \tag{13}$$

where J' is the modified Jacobian matrix.

The condition number of J' is regarded as the local performance index for evaluating the velocity, accuracy and rigidity mapping characteristics between the joint variables and the moving platform. The condition number has the main advantage of being a single number for describing the overall kinematic behavior of a robot. The condition number κ is defined as

$$1 \leq \kappa = \dfrac{\sigma_2}{\sigma_1} \leq \infty \tag{14}$$

where σ_1 and σ_2 are the minimum and maximum singular values of the Jacobian matrix associated with a given posture.

The condition number κ is configuration-dependent. Its reciprocal $1/\kappa$ ($0 \leq 1/\kappa \leq 1$) is a local performance index that is referred to as the local conditioning index. Considering that κ varies with the configuration of the mechanism, a global performance index is used as the performance measure in the optimal kinematic design, namely

$$\eta = \frac{\int_W 1/\kappa \, dW}{\int_W dW} \tag{15}$$

where W is the workspace.

4. Dynamic Model

Since the dynamic performance is considered in the optimum design, the dynamic model should be derived. Here, the virtual work principle is employed to derive the dynamic model. The partial velocity and partial angular velocity matrices, which are used in dynamic modeling, should be first determined. Then, the inertia force and torque of each moving part are computed. Finally, the dynamic model can be obtained.

4.1. Partial Velocity Matrix and Partial Angular Velocity Matrix

For a parallel manipulator with n DOF, n generalized coordinates are needed to specify the system completely. Based on the Jacobian matrix, the partial velocity matrix [20] can be expressed as

$$\boldsymbol{H}_i = \begin{bmatrix} \dfrac{\partial \dot{q}_1}{\partial v} & \dfrac{\partial \dot{q}_2}{\partial v} & \cdots & \dfrac{\partial \dot{q}_n}{\partial v} \end{bmatrix}^T \tag{16}$$

where $\dot{q}_n$ is the velocity of the nth active joint, and v is the end-effector velocity.

Accordingly, the partial angular velocity matrix can be expressed as

$$\boldsymbol{G}_i = \begin{bmatrix} \dfrac{\partial \omega_1}{\partial v} & \dfrac{\partial \omega_2}{\partial v} & \cdots & \dfrac{\partial \omega_n}{\partial v} \end{bmatrix}^T \tag{17}$$

where ω_n is the angular velocity of the nth link.

In order to compute the partial angular velocity matrix, the pivotal point should be determined. For the 3-DOF parallel manipulator, point A_i is selected as the pivotal point of slider $A_i D_i$ and the link $A_i B_i$. Points D_i, C_i, G_i and O' are the pivotal points of the upper part of link $D_i C_i$, the lower part of link $D_i C_i$, the counterweight and the moving platform.

While the slider has only translational capability, the partial velocity and partial angular velocity matrix can be written as

$$H_{i1} = e_2 J_i, \quad G_{i1} = 0 \tag{18}$$

Since the counterweight is connected to the slider and its velocity is the negative of that of the slider, its partial velocity and partial angular velocity matrix can be expressed as

$$H_{i2} = -H_{i1}, \quad G_{i2} = 0 \tag{19}$$

Based on Eq. (7) and (8), the partial velocity and partial angular velocity matrix of link $A_i B_i$ are given by

$$H_{i3} = H_{i1}, \quad G_{i3} = \frac{e_1^T T_{Bi}}{l_i e_1^T E s_i} \tag{20}$$

For the upper part of link $D_i C_i$, the partial velocity and partial angular velocity matrix can be got based on Eq. (7) and (10), as

$$H_{i4} = H_{i1}, \quad G_{i4} = s_{i+2}^T E^T (T_{Ci} - e_2 J_i)/l_{i+2} \tag{21}$$

The partial velocity and partial angular velocity matrix of the lower part of link $D_i C_i$ can be expressed as

$$H_{i5} = T_{Ci}, \quad G_{i5} = G_{i4} \tag{22}$$

The partial velocity and partial angular velocity matrix of the moving platform are given by

$$H_N = [I_2 \quad 0], \quad G_N = [0 \quad 0 \quad 1] \tag{23}$$

4.2. Force Transmission on a Rigid Body

In general, the inertia force and torque about the mass center is easy to be derived by using the Newton-Euler equation [21]. It is complex to obtain them about a random point. In this section, the rule of force transmission on a rigid body is given [22]. Let a coordinate system $\mathcal{E}$ be attached to the mass center C of a body. In general, it is easy to compute the inertia force at the mass center of a rigid body. If the position of the mass center is not known, we can choose an arbitrary coordinate system $\mathcal{A}$ attached to point A of the body and express the inertia force relatively to this coordinate system. A wrench F_C applied to the mass center C and described with respect to coordinate system $\mathcal{E}$ on a rigid body gives the following contribution at point A with respect to coordinate system $\mathcal{A}$

$${}^{\mathcal{A}}F_A = J_r^{\mathrm{T}}\, {}^{e}F_C \tag{24}$$

where $J_r^{\mathrm{T}} = \begin{bmatrix} {}^{\mathcal{A}}_{e}R & 0 \\ {}^{\mathcal{A}}\hat{r}\, {}^{\mathcal{A}}_{e}R & {}^{\mathcal{A}}_{e}R \end{bmatrix}$, $\hat{r}$ is a skew symmetric matrix corresponding to the cross-product with the vector r from point A to C, ${}^{\mathcal{A}}_{e}R$ is the rotation matrix between coordinate systems $\mathcal{A}$ and $\mathcal{E}$.

The inertia force ${}^{e}F_C$ acting at the mass center of the body is calculated using the following equation

$${}^{e}F_C = \begin{bmatrix} m\,{}^{e}a_C \\ {}^{e}I_C\, {}^{e}\dot{\omega}_C + {}^{e}\omega_C \times {}^{e}I_C\, {}^{e}\omega_C \end{bmatrix} \tag{25}$$

where m is the mass of the rigid body, ${}^{e}a_C$ is the acceleration of the body with respect to coordinate system $\mathcal{E}$, ${}^{e}\omega_C$ is the angular velocity of the body, and ${}^{e}I_C$ is the moment of inertia of the body about the mass center.

Given the acceleration a_A of point A on the body, the acceleration a_C of the mass center can be expressed as

$${}^{\mathcal{A}}a_C = {}^{\mathcal{A}}a_A + {}^{\mathcal{A}}\omega_A \times \left({}^{\mathcal{A}}\omega_A \times {}^{\mathcal{A}}r\right) + {}^{\mathcal{A}}\dot{\omega}_A \times {}^{\mathcal{A}}r \tag{26}$$

Thus, the following equation can be obtained

$${}^{\mathcal{A}}F_A = \begin{bmatrix} m\left({}^{\mathcal{A}}a_A + {}^{\mathcal{A}}\omega_A\, {}^{\mathcal{A}}\omega_A\, {}^{\mathcal{A}}r + {}^{\mathcal{A}}\dot{\omega}_A\, {}^{\mathcal{A}}r\right) \\ m\,{}^{\mathcal{A}}r\,{}^{\mathcal{A}}a_A + {}^{\mathcal{A}}I_A\,{}^{\mathcal{A}}\dot{\omega}_A + {}^{\mathcal{A}}\omega_A\,{}^{\mathcal{A}}I_A\,{}^{\mathcal{A}}\omega_A \end{bmatrix} \tag{27}$$

where ${}^{\mathcal{A}}I_A$ is the moment of inertia of the body about point A.

The wrench ${}^{\mathcal{A}}F_A$ can finally be calculated in the base coordinate system $O-XY$ by

$${}^{O}F_A = \begin{bmatrix} {}^{O}_{\mathcal{A}}R & 0 \\ 0 & {}^{O}_{\mathcal{A}}R \end{bmatrix} {}^{\mathcal{A}}F_A \tag{28}$$

where ${}^{O}_{\mathcal{A}}R$ is the rotation matrix between coordinate systems $O-XY$ and $\mathcal{A}$.

4.3. Inertia Force and Torque

Taking the time derivative of Eq. (5) and (6) leads to

$$\ddot{q}_i e_2 + l_i \ddot{\beta}_i Es_i - l_i \dot{\beta}_i^2 s_i = \ddot{r} + \ddot{\alpha} ERr'_{Bi} - \dot{\alpha}^2 Rr'_{Bi} \qquad (29)$$

$$\ddot{q}_i e_2 + \ddot{l}_{i+2} s_{i+2} + 2\dot{l}_{i+2} \dot{\phi}_i Es_{i+2} + l_{i+2} \ddot{\phi}_i Es_{i+2} - l_{i+2} \dot{\phi}_i^2 s_{i+2} = \ddot{r} + \ddot{\alpha} ERr'_{Ci} - \dot{\alpha}^2 Rr'_{Ci} \qquad (30)$$

Multiplying both sides of Eq. (29) by s_i^T and e_1^T leads to

$$\ddot{q}_i = Q_i \ddot{p} + \tilde{\alpha}_i \qquad (31)$$

$$\ddot{\beta}_i = W_i \ddot{p} + \tilde{\varepsilon}_i \qquad (32)$$

where $Q_i = J_i$,

$$\tilde{\alpha}_i = \frac{s_i^T (l_i \dot{\beta}_i^2 s_i - \dot{\alpha}^2 Rr'_{Bi})}{s_i^T e_2},$$

$$W_i = \frac{e_1^T T_{Bi}}{l_i e_1^T Es_i},$$

$$\tilde{\varepsilon}_i = \frac{e_1^T (l_i \dot{\beta}_i^2 s_i - \dot{\alpha}^2 Rr'_{Bi})}{l_i e_1^T Es_i^T}.$$

Multiplying both sides of Eq. (30) by s_i^T and e_1^T leads to

$$\ddot{l}_{i+2} = Q_{i+2} \ddot{p} + \tilde{\alpha}_{i+2} \qquad (33)$$

$$\ddot{\phi}_i = W_{i+2} \ddot{p} + \tilde{\varepsilon}_{i+2} \qquad (34)$$

where $Q_{i+2} = J_{i+2}$,

$$\tilde{\alpha}_{i+2} = s_{i+2}^T (l_{i+2} \dot{\phi}_i^2 s_{i+2} - \dot{\alpha}^2 Rr'_{Ci} - \tilde{\alpha}_i e_2),$$

$$W_{i+2} = s_{i+2}^T E^T (T_{Ci} - e_2 Q_i)/l_{i+2},$$

$$\tilde{\varepsilon}_{i+2} = -\frac{s_{i+2}^T E^T (\dot{\alpha}^2 Rr'_{Ci} + \tilde{\alpha}_i e_2) + 2\dot{l}_{i+2} \dot{\phi}_i}{l_{i+2}}.$$

The inertia force and torque of each moving part about the pivotal point can be determined utilizing the Newton-Euler formulation. m_{i1}, m_{i2}, m_{i3}, m_{i4}, m_{i5} and m_N denote the masses of the slider, counterweight, link $A_i B_i$, the upper part of link $D_i C_i$, the lower part of link D_iC_i and the moving platform, and I_{i1}, I_{i2}, I_{i3}, I_{i4}, I_{i5} and I_N are their inertia matrices about the corresponding pivotal points, and the gravitational acceleration vector is $\boldsymbol{g} = [0 \ -9.8]^T \ m/s^2$.

The inertia force and torque of slider A_iD_i about point A_i can be expressed as

$$\boldsymbol{F}_{i1} = -m_{i1}(\ddot{q}_i\boldsymbol{e}_2 - \boldsymbol{g}) = \boldsymbol{T}_{i1}\ddot{\boldsymbol{p}} + \widetilde{\boldsymbol{F}}_{i1} \tag{35}$$

$$\boldsymbol{M}_{i1} = \boldsymbol{S}_{i1}\ddot{\boldsymbol{p}} \tag{36}$$

where $\boldsymbol{T}_{i1} = -m_{i1}\boldsymbol{e}_2\boldsymbol{Q}_i$,

$\widetilde{\boldsymbol{F}}_{i1} = m_{i1}(\boldsymbol{g} - \widetilde{\alpha}_i\boldsymbol{e}_2)$,

$\boldsymbol{S}_{i1} = \boldsymbol{0}$.

The inertia force and torque of the counterweight about point G_i can be written as

$$\boldsymbol{F}_{i2} = -m_{i2}(-\ddot{q}_i\boldsymbol{e}_2 - \boldsymbol{g}) = \boldsymbol{T}_{i2}\ddot{\boldsymbol{p}} + \widetilde{\boldsymbol{F}}_{i2} \tag{37}$$

$$\boldsymbol{M}_{i2} = \boldsymbol{S}_{i2}\ddot{\boldsymbol{p}} \tag{38}$$

Where $\boldsymbol{T}_{i2} = m_{i2}\boldsymbol{e}_2\boldsymbol{Q}_i$,

$\widetilde{\boldsymbol{F}}_{i2} = m_{i2}(\boldsymbol{g} + \widetilde{\alpha}_i\boldsymbol{e}_2)$,

$\boldsymbol{S}_{i2} = \boldsymbol{0}$.

The inertia force and torque of link A_iB_i about point A_i can be expressed as

$$\boldsymbol{F}_{i3} = -m_{i3}(\ddot{q}_i\boldsymbol{e}_2 + n_i\ddot{\beta}_i\boldsymbol{E}\boldsymbol{s}_i - n_i\dot{\beta}_i^2\boldsymbol{s}_i - \boldsymbol{g}) = \boldsymbol{T}_{i3}\ddot{\boldsymbol{p}} + \widetilde{\boldsymbol{F}}_{i3} \tag{39}$$

$$\boldsymbol{M}_{i3} = -I_{i3}\ddot{\beta}_i + n_i\boldsymbol{s}_i^T\boldsymbol{E}\boldsymbol{F}_{i3} = \boldsymbol{S}_{i3}\ddot{\boldsymbol{p}} + \widetilde{\boldsymbol{M}}_{i3} \tag{40}$$

where n_i is the distance from the mass center of link A_iB_i to point A_i, $\boldsymbol{T}_{i3} = -m_{i3}(\boldsymbol{e}_2\boldsymbol{Q}_i + n_i\boldsymbol{E}\boldsymbol{s}_i\boldsymbol{W}_i)$,

$\widetilde{\boldsymbol{F}}_{i3} = m_{i3}(n_i\dot{\beta}_i^2\boldsymbol{s}_i + \boldsymbol{g} - \widetilde{\alpha}_i\boldsymbol{e}_2 - n_i\widetilde{\varepsilon}_i\boldsymbol{E}\boldsymbol{s}_i)$,

$\boldsymbol{S}_{i3} = -I_{i3}\boldsymbol{W}_i - m_{i3}n_i\boldsymbol{s}_i^T\boldsymbol{E}(\boldsymbol{e}_2\boldsymbol{Q}_i + n_i\boldsymbol{E}\boldsymbol{s}_i\boldsymbol{W}_i)$,

$$\widetilde{M}_{i3} = n_i s_i^T E \widetilde{F}_{i3} - I_{i3} \widetilde{\varepsilon}_i .$$

For the upper part of link $D_i C_i$, its inertia force and torque about point D_i are determined by

$$F_{i4} = -m_{i4}(\ddot{q}_i e_2 + t_i \ddot{\phi}_i E s_{i+2} - t_i \dot{\phi}_i^2 s_{i+2} - g) = T_{i4} \ddot{p} + \widetilde{F}_{i4} \tag{41}$$

$$M_{i4} = -I_{i4} \ddot{\phi}_i + t_i s_{i+2}^T E \widetilde{F}_{i4} = S_{i4} \ddot{p} + \widetilde{M}_{i4} \tag{42}$$

where t_i is the distance from the mass center of link $D_i C_i$ to point D_i, $T_{i4} = -m_{i4}(e_2 Q_i + t_i E s_{i+2} W_{i+2})$,

$$\widetilde{F}_{i4} = m_{i4}(t_i \dot{\phi}_i^2 s_{i+2} + g - \widetilde{\alpha}_i e_2 - t_i \widetilde{\varepsilon}_{i+2} E s_{i+2}),$$

$$S_{i4} = -I_{i4} W_{i+2} - m_{i4} t_i s_{i+2}^T E(e_2 Q_i + t_i E s_{i+2} W_{i+2}),$$

$$\widetilde{M}_{i4} = t_i s_{i+2}^T E \widetilde{F}_{i4} - I_{i4} \widetilde{\varepsilon}_{i+2} .$$

For the lower part of link $D_i C_i$, its inertia force and torque about point C_i are determined by

$$F_{i5} = -m_{i5}(T_{Ci} \ddot{p} - \dot{\alpha}^2 R r'_{Ci} - u_i \ddot{\phi}_i E s_{i+2} + u_i \dot{\phi}_i^2 s_{i+2} - g) = T_{i5} \ddot{p} + \widetilde{F}_{i5} \tag{43}$$

$$M_{i5} = -I_{i5} \ddot{\phi}_i - u_i s_{i+2}^T E F_{i5} = S_{i5} \ddot{p} + \widetilde{M}_{i5} \tag{44}$$

where u_i is the distance from the mass center of link $D_i C_i$ to point C_i, $T_{i5} = -m_{i5}(T_{Ci} - u_i E s_{i+2} W_{i+2})$,

$$\widetilde{F}_{i5} = m_{i5}(\dot{\alpha}^2 R r'_{Ci} + g - u_i \dot{\phi}_i^2 s_{i+2} + u_i \widetilde{\varepsilon}_{i+2} E s_{i+2}),$$

$$S_{i5} = -I_{i5} W_{i+2} + m_{i5} u_i s_{i+2}^T E(T_{Ci} + u_i E s_{i+2} W_{i+2}),$$

$$\widetilde{M}_{i5} = -u_i s_{i+2}^T E \widetilde{F}_{i5} - I_{i5} \widetilde{\varepsilon}_{i+2} .$$

The inertia force and torque of the moving platform about point O' are given as

$$F_N = -m_N [I_2 \quad 0] \ddot{p} = T_N \ddot{p} \tag{45}$$

$$M_N = I_N \ddot{\alpha} = S_N \ddot{p} \tag{46}$$

where $T_N = -m_N[I_2 \quad 0]$ and $S_N = [0 \quad 0 \quad I_N]^T$.

4.4. Dynamic Model

Based on the virtual work principle, the dynamic formulation of the redundantly actuated parallel manipulator can be expressed as

$$J^T \tau + \sum_{i=1}^{2}\sum_{j=1}^{5}[H_{ij}^T \quad G_{ij}^T]\begin{bmatrix}F_{ij}\\M_{ij}\end{bmatrix} + [H_N^T \quad G_N^T]\begin{bmatrix}F_N\\M_N\end{bmatrix} = 0 \qquad (47)$$

where $\tau = [F_1 \quad F_2 \quad F_3 \quad F_4]^T$, and F_1, F_2, F_3 and F_4 are the driving forces that act on sliders A_1D_1, A_2D_2 and the extendible links D_1C_1, D_2C_2.

Eq. (47) can be rewritten as

$$J^T \tau + M\ddot{P} + \tilde{M} = 0 \qquad (48)$$

where $M = \sum_{i=1}^{2}\sum_{j=1}^{5}[H_{ij}^T \quad G_{ij}^T]\begin{bmatrix}T_{ij}\\S_{ij}\end{bmatrix} + [H_N^T \quad G_N^T]\begin{bmatrix}T_N\\S_N\end{bmatrix}$,

$\tilde{M} = \sum_{i=1}^{2}\sum_{j=1}^{5}[H_{ij}^T \quad G_{ij}^T]\begin{bmatrix}\tilde{F}_{ij}\\\tilde{M}_{ij}\end{bmatrix}$, $\tilde{M}$ consists of the centrifugal, Coriolis and gravitational force.

5. DYNAMIC PERFORMANCE INDEX

Both the DME and GIE are based on the relationship between the generalized acceleration of the end effector and the generalized inertia torques of the joints. As addressed in [8], the dynamic manipulability ellipsoid for redundant manipulators is proposed by resorting to the unweighted pseudoinverse of the Jocabian matrix. Thus, Eq. (48) can be rewritten in a unified form by neglecting $\tilde{M}$, lead to

$$\tau \approx J^T(JJ^T)^{-1} M\ddot{P} \qquad (49)$$

where $J^T(JJ^T)^{-1} M$ is the generalized inertia matrix.

Based on the GIE, the moving platform can easily accelerate in the direction of major axis of the ellipsoid and hardly in the direction of minor axis. The lengths of the principal axes represent the maximum and minimum singular values of the inertia matrix, and the difference between them stands for the anisotropy of the accelerating performance, which is isotropic when the lengths of the principal axes are the same.

$$1 \leq \kappa_D = \frac{\sigma_{D2}}{\sigma_{D1}} \leq \infty \qquad (50)$$

where σ_{D1} and σ_{D2} are the minimum and maximum singular values of the generalized inertia matrix, respectively.

The dynamic dexterity can be evaluated by κ_D. However, it is a local performance index. To obtain a measure of the global behavior of the manipulator, dynamic condition number, similar to that introduced in [23,24], a global dynamic conditioning index η_D is proposed:

$$\eta_D = \frac{\int_W 1/\kappa_D \, dW}{\int_W dW} \qquad (51)$$

The index defined in Eq. (51) is to maximize over the space of manipulator parameters. Thus, the closer to 1 is η_D, the better is the overall dynamic dexterity of the manipulator.

As an example to investigate the dynamic dexterity, the geometrical parameters and inertial parameters are given in Table 1 and 2, respectively. The geometrical parameters are obtained based on kinematic optimum design in reference [25].

Table 1. geometrical parameters

Parameter	Value
d (m)	1.17
l_1 (m)	1.15
l_2 (m)	1.15
l_5 (m)	0.25
l_6 (m)	0.25

Table 2. inertial parameters

	$i=1$	$i=2$
m_N (kg)	150	150
m_{i1} (kg)	120	120
m_{i2} (kg)	495	495
m_{i3} (kg)	220	220
m_{i4} (kg)	60	60
m_{i5} (kg)	20	20
I_{i3} (kg.m^2)	105.6	105.6
I_{i4} (kg.m^2)	7.2	7.2
I_{i5} (kg.m^2)	4.27	4.27
I_N (kg.m^2)	5.45	5.45

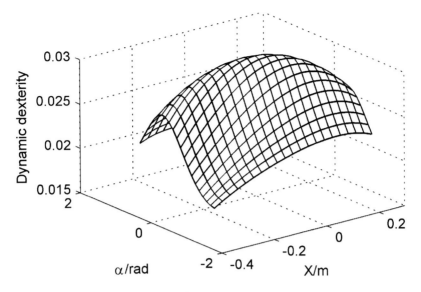

Figure 3. Dynamic dexterity of the inertia matrix.

From Eq. (7), (9) and (48), one may see that the dynamic dexterity κ_D is independent of y in the task workspace. Thus, in the following study, the effect of y on κ_D and η_D is not considered. Based on the parameters shown in Tables 1 and 2, $1/\kappa_D$ of the redundant parallel manipulator in the workspace with $-0.3m \leq x \leq 0.3m$ and $-60° \leq \alpha \leq 60°$ is shown in Figure 3. One may see that $1/\kappa_D$ has a larger value in the center where $x = 0$ and $\alpha = 0°$.

6. OPTIMUM DESIGN BASED ON KINEMATICS AND DYNAMICS

In the optimum design, the task workspace W of the redundantly actuated parallel manipulator is designed as a rectangle of $b = 970$mm in width and $h = 630$mm in height. The width d between two columns can be expressed as

$$d = b + 2l_1 \sin \beta_{1\min} \tag{52}$$

where $\beta_{1\min}$ is the minimum value of β_1, and $\beta_{1\min}$ is given by the designer in terms of the minimum stiffness of the manipulator and mechanism interference.

In reference [25], it is specified that $\beta_{1\min} = 5°$ and the kinematic optimization is carried out. The optimization results are: $d = 1170$ mm, $l_2 = l_3 = 1150$mm, and $l_5 = 250$mm.

As mentioned in Section I, the dynamic characteristics should be considered in the optimum design to realize the high dynamics. Here, the dynamic performance index η_D is considered in the process of optimum design. Besides the geometrical parameters, the inertial parameters are also needed to compute η_D. It is assumed that the material is evenly distributed and the mass can be computed through multiplying density by volume. If the cross-section is given, only the geometrical parameters are unknown and should be determined in the dynamic performance index. The cross-section parameters of links can be optimized by stiffness analysis. Here, the cross section of constant-length link is a rectangle and the cross section of extendible link is circular.

Based on the global dynamic dexterity index η_D and the results of kinematic optimum design, the geometrical parameters of the manipulator are optimized. The value of l_1 and l_2 are equal in practical application, l_5 and l_6 are the same. It is defined that $1m \leq l_1 \leq 1.5m$, $0.2m \leq l_5 \leq 0.5m$. The relationship among η_D, d, l_1 and l_5 are shown in Figures (4), (5) and (6). The smaller d is, the better the dynamic dexterity η_D is. The larger l_1 is, the better the dynamic dexterity η_D is. Since the task workspace W is specified in the optimum design, the width b of W can not be changed in the optimum design by considering the dynamics. From Figure 1, one may see that l_1 shall be larger than b. Thus, $d = b + 2l_1 \sin \beta_{1\min} > 0.97 + 2 \times 0.97 \times \sin 5^0 = 1.14$ m. Since a smaller d is helpful for the improvement of dynamic dexterity, in the optimum design based on dynamics, d is still equal to 1.17m which is obtained by kinematic optimal design.

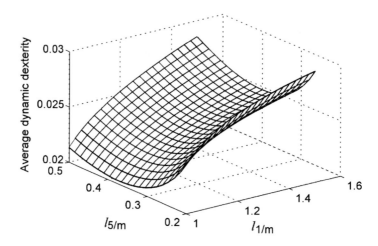

Figure 4. relationship among η_D, l_1 and l_5.

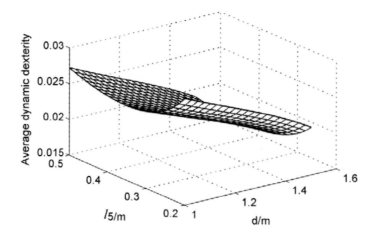

Figure 5. relationship among η_D, d and l_5.

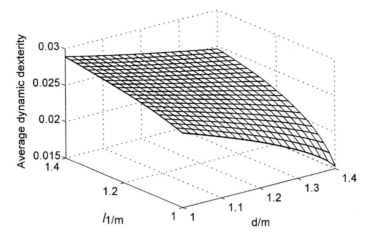

Figure 6. relationship among η_D, d and l_1.

Figure 7 shows the contour map of η_D. One may see that l_1 is inversely proportional to η_D. When l_5 ranges between 0.25 and 0.45, η_D has a smaller value. When l_5 range from 0.25 to 0.2, and from 0.45 to 0.5, η_D has a larger value. In order to maintain the better kinematic performance of the manipulator, the values of l_1 and l_5 should approach the optimal results from kinematic optimum design. Therefore, the value of l_5 should determined in the range of 0.2m to 0.25m, and cannot range from 0.45m to 0.5m. It is defined that l_5 =0.21m and η_D=0.028. Based on Figure 7, it can be concluded that l_1=1.28m. $1/\kappa_D$ of the re-designed parallel manipulator is shown in Figure 8. One may see that the manipulator has a better dynamic dexterity in the center of the task workspace and the maximum value of $1/\kappa_D$ is about 0.035. Thus, the manipulator has the desirable dynamic performance. Further, from Figures 3 and 8, it can be concluded that the designed manipulator based on kinematics and dynamics has a better dynamic performance.

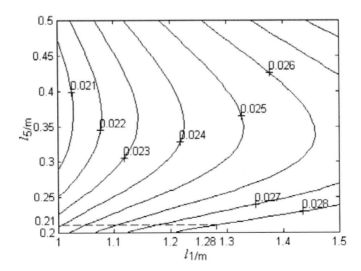

Figure 7. Counter map of η_D over the parameter space.

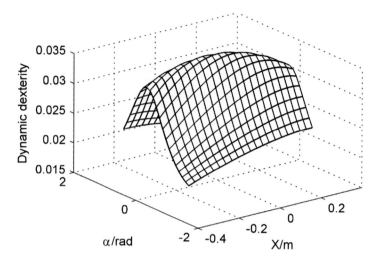

Figure 8. Dynamic dexterity of the improved mechanism.

In order to investigate the performance of the designed manipulator with considering kinematics and dynamics, the distribution of κ in the workspace is given in Figure 9. One may see that the condition number distribution is symmetrical. The maximum and minimum values of κ are about 1.5 and 2.5. In order to compare the manipulator designed based on kinematics and dynamics and that designed based on kinematics, the distribution of κ of the one designed based on kinematics is also given in Figure 10. It can be seen that the distribution of κ in the workspace is also symmetrical and the minimum and maximum values are about 1 and 3. Thus, the manipulator designed based on kinematics has a better kinematic performance in the center of the task workspace due to the smaller κ. However, in the task workspace, the kinematic performance of the original manipulator changes abruptly for the larger difference between the maximum value and minimum value of κ.

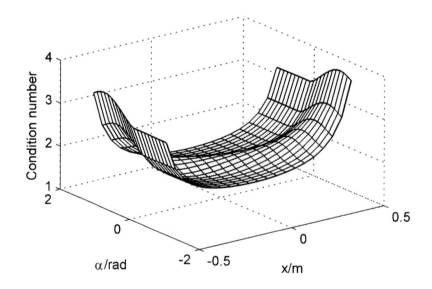

Figure 9. κ of the manipulator with kinematic and dynamic optimum design.

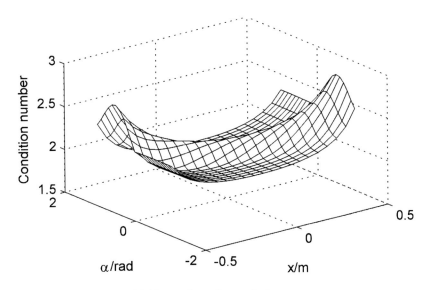

Figure 10. κ of the manipulator with kinematic optimum design.

Therefore, the re-designed parallel manipulator with considering kinematics and dynamics still has a good kinematic performance. The method proposed in the chapter would be practical for the optimum design of parallel manipulators.

CONCLUSION

In the chapter, a new method for optimum design of parallel manipulators is proposed by considering both kinematics and dynamics. This method is divided into two steps: 1) kinematic optimum design is carried out; 2) optimum design based on dynamic performance index is performed. Compared with other optimum design methods, the proposed method not only considers the kinematic performance, but also involves the dynamic performance in the kinematic design phase. Although the kinematic performance may be debased by using the method proposed in the chapter, the dynamic performance is improved. Since the dynamic performance can be considered in the optimum design before the prototype is built, the results of this chapter are helpful for the realization of the high dynamics of parallel manipulators.

REFERENCES

[1] Siciliano, B. (1999). The Tricept robot: inverse kinematics, manipulability analysis and closed-loop direct kinematics algorithm. *Robotica*, Vol.17, 437-445.

[2] Pierrot, F., Nabat, V., Company, O., Krut, S., and Poignet, P. (2009). Optimal design of a 4-DOF parallel manipulator: from academia to industry. *IEEE Transactions on Robotics*, vol. 25, 213-224.

[3] Gao, F, Liu, X.-J., Chen, X. (2001). The relationships between the shapes of the workspaces and the link lengths of 3-DOF symmetrical planar parallel. *Mechanism and Machine Theory*, vol. 36, 205-220.

[4] Liu, X.-J., and Wang, J.S. (2007). A new methodology for optimal kinematic design of parallel mechanisms. *Mechanism and Machine Theory*, vol. 42, 1210-1224.

[5] Liu, X.-J., Guan, L.W., Wang, J.S. (2007). Kinematics and closed optimal design of a kind of PRRRP parallel manipulator.. *Journal of Mechanical Design*, vol. 129, 558-563.

[6] Liu, X.-J., Wang, J.S., J. Kim. (2006). Determination of the link lengths for a spatial 3-DOF parallel manipulator. *Journal of Mechanical Design*, vol. 128, 365-373.

[7] Cervantes-Sa'nchez, J.J., Herna'ndez-Rodrıguez, J.C., and Angeles, J. (2001). On the kinematic design of the 5R planar, symmetric manipulator. *Mechanism and Machine Theory*, vol. 36, 1301–1313.

[8] Yoshikawa, T. (1985). Dynamic manipulability of robot manipulators. *Journal of Robotic Systems*, vol. 2, 113-124.

[9] Yoshikawa, T. (1985). Manipulability of robotic mechanisms. *The International Journal of Robotics Research*, vol. 4, 3-9.

[10] Asada, H. (1983). A geometrical representation of manipulator dynamics and its application to arm design. *Journal of Dynamic Systems, Measurement, and Control*, vol. 105, 131-135.

[11] Graettinger, T. J. and Krogh, B. H. (1988). The acceleration radius: a global performance measure for robotic manipulators. *IEEE Journal on Robotics and Automation*, vol. 4, 60-69.

[12] Hashimoto, R. (1985). Harmonic mean type manipulability index of robotic arms. *Transactions of the Society of Instrument and Control Engineers*, vol. 21, 1351-1353.

[13] Li, M., Huang, T., Mei, J., Zhao, X., Chetwynd, D. G., and Hu, S. J. (2005). Dynamic formulation and performance comparison of the 3-DOF modules of two reconfigurable PKM-the Tricept and the TriVariant. *Journal of Mechanical Design*, vol. 127, 1129-1136.

[14] Zhao, Y., and Gao, F. (2009). Dynamic performance comparison of the 8PSS redundant parallel manipulator and its non-redundant counterpart-the 6PSS parallel manipulator. *Mechanism and Machine Theory*, vol. 44, 991-1008.

[15] Mansouri, I., and Ouali, M. (2009). A new homogeneous manipulability measure of robot manipulators, based on power concept. *Mechatronics*, vol.19, 927-944.

[16] Nokleby, S. B., Fisher, R., Podhorodeski, R.P., and Firmani, F. (2005). Force capabilities of redundantly-actuated parallel manipulators. *Mechanism and Machine Theory*, vol. 40, 578-599.

[17] Chiacchio, P. and Concilio, M. (1998). The Dynamic Manipulability Ellipsoid for Redundant Manipulators, In: Proceedings of IEEE International Conference on Robotics and Automation, Leuven, Belgium, 1998, pp. 95-100.

[18] Asada, H. Dynamic analysis and design of robot manipulators using inertia ellipsoids In: Proceedings of IEEE International Conference on Robotics, Atlanta, GA, USA, 1984, pp. 94-102.

[19] Wu, J., Wang, J.S., Wang, L.P., and Li, T.M. (2009). Dynamics and control of a planar 3-DOF parallel manipulator with actuation redundancy. *Mechanism and Machine Theory*, vol.44, 835-849.

[20] Zhang, C.D. and Song S.M. (1993). An efficient method for inverse dynamics of manipulate ors based on the virtual work principle. *Journal of Robotic Systems*, vol. 10, 605-627.

[21] Codourey, A. Dynamic modeling of parallel robots for computed-torque control implementation. *International Journal of Robotics Research*, vol. 17, 1325-1336.

[22] Codourey, A. and Burdet, E. A body-oriented method for finding a linear form of the dynamic equation of fully parallel robots. Proceedings of the IEEE International Conference on Robotics and Automation, Albuquerque, New Mexico, 1997, 1612-1618.

[23] Wu, J., Wang, J.S., Li, T.M., and Guan, L.W. (2008). Dynamic dexterity of a planar 2-DOF parallel manipulator in a hybrid machine tool. *Robotica*, vol. 26, 93-98.

[24] Sho, H., Wang, L.P., Guan, L.W., Wu, J. Dynamic manipulability and optimization of a redundant three DOF planar parallel manipulator. Proceedings of the ASME International Conference on Reconfigurable Mechanisms and Robots, London, United Kingdom, 2009, 302-308.

[25] Wu, J., Wang, J.S., and Wang, L. (2008). Optimal kinematic design and application of a redundantly actuated 3DOF planar parallel manipulator. *Journal of Mechanical Design*, vol.130, 0545031-0545035.

In: Machine Tools: Design, Reliability and Safety
Editor: Scott P. Anderson, pp. 175-185

ISBN 978-1-61209-144-0
© 2011 Nova Science Publishers, Inc.

Chapter 6

SITE CHARACTERIZATION MODEL USING MACHINE LEARNING

Sarat Das[1],, Pijush Samui[2],♦ and D. P. Kothari[3]•*

[1] Department of Civil Engineering,
National Institute of Technology Rourkela, India
[2] Centre for Disaster Mitigation and Management.
VIT University, Vellore, India
[3] VIT University, Vellore, India

ABSTRACT

This chapter describes two machine learning techniques for developing site characterization model. The ultimate goal of site characterization is to predict the in-situ soil properties at any half-space point for a site based on limited number of tests and data. In three dimensional analysis, the function $N = N(X, Y, Z)$ where X, Y and Z are the coordinates of a point corresponds to Standard Penetration Test(SPT) value(N), is to be approximated with which N value at any half space point in site can be determined. The site is located in the alluvial Gangetic plane (Sahajanpur of Uttar Pradesh, India). The input of machine leaning techniques is X, Y and Z. The output of machine learning techniques is N. The first machine learning technique uses generalized regression neural network (GRNN) that are trained with suitable spread(s) to predict N value. The second machine learning technique uses Least Square Support Vector Machine (LSSVM), is a statistical learning theory which adopts a least squares linear system as a loss function instead of the quadratic program in original support vector machine (SVM). Here, LSSVM has been used as a regression technique. The developed LSSVM model has been used to compute error bar of the predicted data. An equation has been also developed for the prediction of N value based on the developed LSSVM model. A comparative study between the two developed machine learning techniques has been presented in this chapter. This chapter shows that the developed LSSVM model is better than GRNN.

---

* E-mail:saratdas@rediffmail.com; sarat@nitrkl.ac.in.
♦ E-mail: pijush.phd@gmail.com.
• E-mail: vc@vit.ac.in.

Keywords: Machine Learning; Least Square Support Vector Machine; Generalized Regression Neural Network; Site Characterization

INTRODUCTION

One of the most important steps in geotechnical engineering is site characterization. The basic objective of site characterization is to provide sufficient and reliable information and data of the site to a level compatible and consistent with the needs and requirements of the project. In situ tests based on Standard Penetration Test (SPT), Cone Penetration Test (CPT) and shear wave velocity method are popular among geotechnical engineering. These tests are time consuming and expensive. Modelling of spatial variation of soil properties based on limited finite number of in situ test data is an imperative task in probabilistic site characterization. It has been used to design future soil sampling programs for the site and to specify the soil stratification. It is never possible to know the geotechnical properties at every location beneath an actual site because, in order to do so, one would need to sample and/or test the entire subsurface profile. So one has to predict geotechnical properties at any point of a site based on a limited number of tests. The prediction of soil property is a difficult task due to uncertainty. Spatial variability, measurement 'noise', measurement and model bias, and statistical error due to limited measurements are the sources of uncertainties. In probabilistic site characterization, random field theory has been used by many researchers in geotechnical engineering (Yaglom, 1962; Lumb, 1975; Vanmarcke, 1977; Tang, 1979; Wu and Wong, 1981; Asaoka and Grivas, 1982; Vanmarcke, 1983; Baecher, 1984; Kulatilake and Miller, 1987; Kulatilake, 1989; Fenton, 1998; Phoon and Kulhawy, 1999; Uzielli et al., 2005). However, the success of random field theory is limited (Juang et al., 2001). Geostatistics and kriging (Matheron, 1963; Journel and Huijbregts, 1978) also have been used to model spatial variation of soil properties (Kulatilake and Ghosh, 1988; Kulatilake, 1989; Soulie et al., 1990; Chiasson et al., 1995; DeGroot, 1996). However, the literature on three dimensional (3D) site characterizations using geostatistics is not available (Juang et al., 2001).

This study uses two machine learning techniques for prediction of SPT value (N) at any point in Sahajanpur (Uttar Pradesh, India). The first machine learning technique uses generalized regression neural network (GRNN) that are trained with suitable spread(s) to predict N value. The second machine learning technique uses Least Square Support Vector Machine (LSSVM), is a statistical learning theory which adopts a least squares linear system as a loss function instead of the quadratic program in original support vector machine (SVM). SPT is one of the most widely used in-situ penetration test designed to provide information on the geotechnical properties of soils. The SPT values (N) can be empirically related to many engineering properties such as unit weight (γ), relative density (D_r), angle of internal friction (ϕ), and undrained compressive strength (q_u) and soil modulus (E_s). The chapter has the following aims:

- To investigate the feasibility of the GRNN and LSSVM model for site characterization.
- To develop an equation for prediction of N value based on the developed LSSVM model
- To compare the performance of developed GRNN and LSVM model.

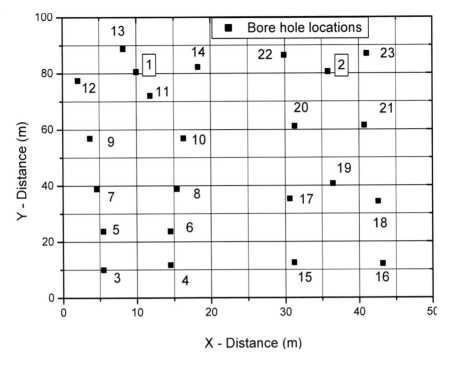

Figure 1. Location of boreholes.

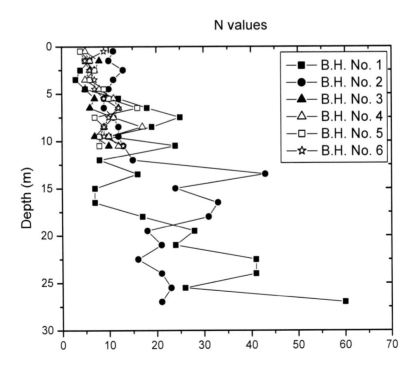

Figure 2. Bore log data showing the variation of SPT (N) value with depth for bore hole number 1 to 6.

GENERAL SITE DESCRIPTION AND GEOTECHNICAL DATA

The considered site is located in the alluvial Gangetic plane (Sahajanpur of Uttar Pradesh, India). Site investigation was carried out to explore the soil strata. Out of these boreholes (Figure 1) two bore holes (Bore hole-1 and Bore hole-2) are of 27m depth and the other bore holes are of 10.5m depth. As the site is a seismic prone area SPT has been conducted at 1.0m interval upto 10.5m and then after at an interval of 1.5m for Bore hole-1 (BH-1) and Bore hole-2 (BH-2). A typical bore log data for 6 bore holes are shown in Figure 2. The bore log data shows a top clay layer extending up to 2.0m underlain by approximately 1.0m of silty sand layer. Thereafter a sand layer with or without kanker extends till 10.0m.

GRNN MODEL

In a GRNN design (Figure 3), hidden layer weights (W_R) are simply the transpose of input vectors from the training set. A Euclidean distance is calculated between an input vector and these weights (Specht, 1991).

$$dist = \left| X - W_R^J \right| \text{ for J=1,Q} \tag{1}$$

Where Q=number of neurons in the hidden layer; X=input vector, dist=Euclidean distance between X and W^J_R. For prediction of N value, X= [X,Y,Z]. Where X, Y and Z are the coordinates of a point corresponds to N. The calculated Euclidean distance is then rescaled by the bias, b:

$$b = 0.8326/s \tag{2}$$

$$n_1 = dist \times b \tag{3}$$

where n_1 = the adjusted distance, and s = the spread.

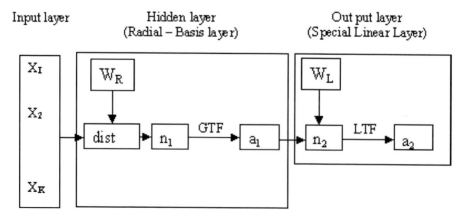

Figure 2. Architecture of GRNN model.

The radial basis output is then the exponential of the negatively adjusted distance having the form:

$$a_1 = e^{\left[-(n_1)^2\right]} \qquad (4)$$

Therefore, if a neuron weight is equal to the input vector, distance between the two is 0 giving an output of 1. This type of neuron gives an output characterizing the closeness between input vectors and weight vectors. The weight matrix size is defined by the size of the training dataset, while the number of neurons is the number of input vectors. The output layer consists of neurons with a linear transfer function, which is:

$$n_2 = W_L \times a_1 \qquad (5)$$

where W_L is the weight matrix in the output layer.

In designing a GRNN model, it is very important to select a proper spread(s).

The main scope of this study is to implement the above methodology for prediction of N value at any point in Sahajanpur. For predicting N in a given space, the three input variables (X,Y,Z) are used for the GRNN model in this study. The data is normalized between 0 to 1. In carrying out the formulation, the data has been divided into two sub-sets: such as

(a) A training dataset: This is required to construct the model. In this study, 234 out of the 334 N values are considered as training dataset.
(b) A testing dataset: This is required to estimate the model performance. In this study, the remaining 100 data is considered as testing dataset.

s is the most important parameter in a GRNN, has been determined using the procedure described below:

(i) Begin the training of the intended GRNN by assuming a small initial s value.
(ii) Evaluate the trained GRNN using the testing data set. This is done simply by comparing the GRNN-predicted N value with the known N value for each case in the testing data set. A root mean squared error (RMSE) of all cases is calculated:

$$\text{RMSE} = \sqrt{\sum_{i=1}^{n} \varepsilon^2 / n} \qquad (6)$$

where, ε is the difference between the known d value and the predicted d value, and n is the number of cases in the testing data set.

(iii) Repeat steps (i) and (ii) above, a number of times using a gradually increased s value. With each adopted s, a GRNN is trained and used to predict N value for each case in the testing data set, and RMSE is then calculated.
(iv) Plot the RMSE against the corresponding s value. The s that yields a minimum RMSE is considered to be the optimum s.

The GRNN model has been developed by using MATLAB.

LSSVM Model

LSSVM models are an alternate formulation of SVM regression (Vapnik and Lerner, 1963) proposed by Suykens et al (2002). Consider a given training set of N data points $\{x_k, y_k\}_{k=1}^N$ with input data $x_k \in R^N$ and output $y_k \in r$ where R^N is the N-dimensional vector space and r is the one-dimensional vector space. The two input variables used for the LSSVM model in this study are latitude and longitude. The output of the LSSVM model is d. So, in this study, $x = [X, Y, Z]$ and $y = N$. In feature space LSSVM models take the form

$$y(x) = w^T \varphi(x) + b \tag{7}$$

where the nonlinear mapping $\varphi(.)$ maps the input data into a higher dimensional feature space; $w \in R^n$; $b \in r$; w = an adjustable weight vector; b = the scalar threshold. In LSSVM for function estimation the following optimization problem is formulated:

$$\text{Minimize:} \frac{1}{2} w^T w + \gamma \frac{1}{2} \sum_{k=1}^N e_k^2$$

Subjected to: $y(x) = w^T \varphi(x_k) + b + e_k$, k=1,...,N. \qquad (8)

The following equation for N prediction has been obtained by solving the above optimization problem (Vapnik, 1998; Smola and Scholkopf, 1998).

$$N = y(x) = \sum_{k=1}^N \alpha_k K(x, x_k) + b \tag{9}$$

The radial basis function has been used in this analysis. The radial basis function is given by

$$K(x_k, x_l) = \exp\left\{-\frac{(x_k - x_l)(x_k - x_l)^T}{2\sigma^2}\right\} \text{ k,l=1,...,N} \tag{10}$$

where σ is the width of radial basis function.

The plausibility of the above proposed models is evaluated using N in the subsurface of Bangalore. In LSVM model, the same training dataset, testing dataset, and normalization technique have been used as used in SVM model. The value of γ and width of radial basis function (σ) value has been chosen by trial and error approach. In the present study, training and testing of LSSVM has been carried out by using MATLAB.

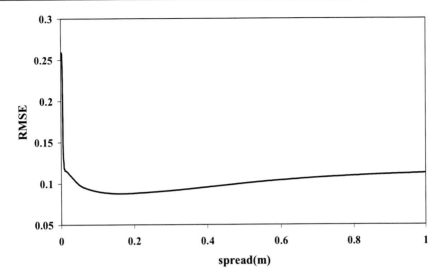

Figure 3. Determination of optimum s for GRNN model.

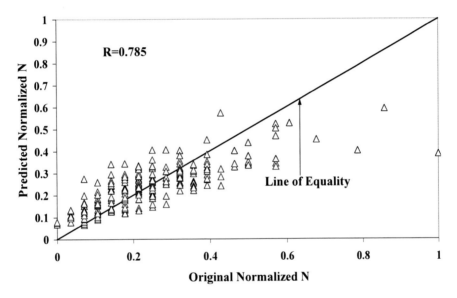

Figure 4. Performance of GRNN model for training dataset.

RESULTS AND DISCUSSION

Coefficient of correlation(R) has used as main criteria to examine the performance of developed model. Figure 3 shows the plot between s and RMSE for GRNN model. From Figure 3, it is clear that 0.1571 is the optimum s.

The performance of training and testing data has been determined by using the value(s=0.1571) of optimum s. Figure 4 and 5 shows the performance of GRNN model for training and testing data respectively. Figure 5 also confirms that the developed GRNN model has capability to predict N at any point in the site. In case of LSSVM model, the design value γ and σ is 100 and 0.4 respectively. Figure 6 and 7 shows the performance of LSSVM model for training and testing dataset respectively. Both GRNN and LSSVM have better performance in the training phase than in the testing phase. The loss of performance with respect to the testing set addresses a machine's susceptibility to overtraining. There is a very marginal reduction in performance on the testing dataset (i.e., there is a difference between machine performance on training and testing) for the GRNN as well as LSSVM model. This relatively small decline of performance of the LSSVM over GRNN model indicates its ability to avoid overtraining, and hence it can be expected to generalize better than GRNN. The following equation can (by putting σ=0.4 and b=0.148 in equation 9) be developed based on LSSVM for the prediction of N at any point in the site.

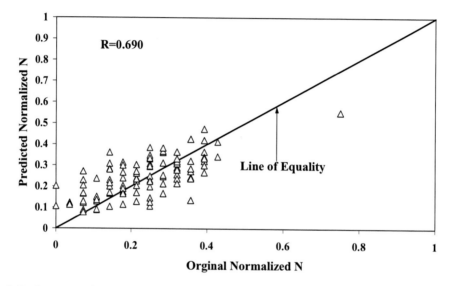

Figure 5. Performance of GRNN model for training dataset.

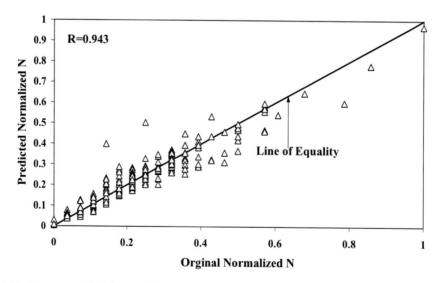

Figure 6. Performance of LSSVM model for training dataset.

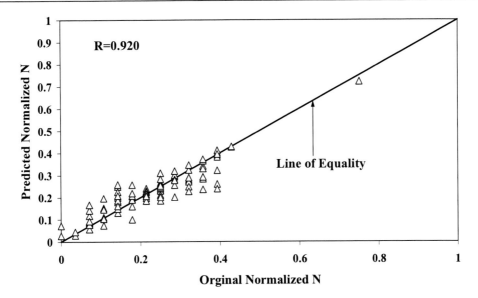

Figure 7. Performance of LSSVM model for testing dataset.

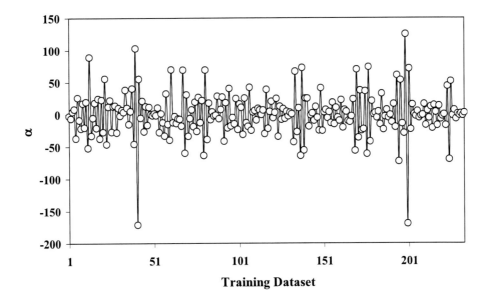

Figure 8. Values of α.

$$N = \sum_{k=1}^{234} \alpha_k \exp\left\{-\frac{(x_i - x)(x_i - x)^T}{0.32}\right\} + 0.148 \qquad (11)$$

Figure 8 shows the value of α.

Table 1. Comparison between LSSVM and GRNN model

X(m)	Y(m)	Z(m)	Actual N	Predicted N by GRNN	Predicted N by LSSVM
12.35	1.3	6.5	12	9	13
6.2	7.6	1.5	4	7	4
12.16	7.25	3.5	10	8	11
18.15	7.12	10	8	11	7
5.6	30.3	9.5	8	10	8

In order to compare between the developed GRNN and LSSVM model, five points have been chosen randomly from the site. The predicted N values of these points are shown in Table 1. It can be seen from the table that the developed LSSVM model outperforms GRNN.

CONCLUSION

This chapter has described two machine learning (GRNN and LSSVM) for prediction of N values in a particular site. The chapter illustrates how GRNN and LSSVM model used in generalising soil properties based on N value. The performance of LSSVM model is better than GRNN model. The developed LSSVM model also gives good generalization capability. User can use the developed equation for prediction of N at any point in the site. The developed models can be used as a quick tool for determination of N value at any point in the site. This study shows that the developed LSSVM is a robust model for site characterization.

REFERENCES

Asaoka, A.; Grivas, D.A. Spatial variability of the undrained strength of clays. *Journal of Geotechnical Engineering.* 1982, 108(5), 743-745.

Baecher, G. B. On estimating auto-covariance of soil properties. Specialty Conference on Probabilistic Mechanics and Structural Reliability, *ASCE.* 1984, 110,214-218.

Baecher, G.B. Geotechnical error analysis. *Trans. Research Record.* 1986, No. 1105, 23-31.

Chiasson, P.; Lafleur, J.; Soulie, M; Law, K.T. Charaterizing spatial variability of clay by geostatistics. *Canadian Geotechnical Journal.* 1995, 32, 1-10.

Degroot, D.J. Analyzing spatial variability of in situ soil properties. ASCE proceedings of uncertainty'96, uncertainty in the geologic environment: from theory to practice, ASCE geotechnical special publications. 1996, 58, 210-38.

Delhomme, J.P. Spatial variability and uncertainty in ground water flow parameters: a geostatistical approach. *Water Resources Research.* 1979, 15(2), 281-290.

Dibike, Y.B.; Velickov, S.; Solomatine, D.; Abbot, M.B. Model Induction with Support Vector machine: Introduction and Application. *Journal of Computing in Civil Engineering.* 2001, 15(3), 208-216.

Fenton, G.A. Random field characterization NGES data. Workshop on Probabilistic Site Characterization at NGES, Seattle, Washington, 1998.

Journel, A.G.; Huijbregts, C.J. Mining Geostatistics; New York Academic Press, NewYork, 1978.

Juang, C.H.; Jiang, T.; Christopher, R.A. Three-dimensional site characterisation: neural network approach. *Geotechnique.* 2001, 51(9), 799-809.

Kulatilake, P.H.S.W. Probabilistic Potentiometric surface mapping. *Journal of Geotechnical Engineering ASCE.* 1989, 115(11), 1569-1587.

Kulatilake, P.H.S.W and Miller, K.M. A scheme for estimating the spatial variation of soil properties in three dimensions. Proc. of the Fifth Int. Conf. On Application of Statistics and Probabilities in Soil and Struct Engrg., Vancouver, BC, Canada; 1987; 669-677.

Lumb. P. Spatial variability of soil properties. Proc. Second Int. Conf. On Application of Statistics and Probability in Soil and struct. Engrg. Aachen, Germany. 1975; 397-421.

Matheron, G. Principles of geostatistics. *Economic Geology.* 1963, 58, 1246-1266.

Phoon, K.K.; Kulhawy, F.H.. Characterization of geotechnical variability. *Canadian Geotechnical Journal.* 1999, 36(4), 612-624.

Scholkopf, B.; Smola, A.J. Learning with kernels: Support Vector Machines, Regularization, Optimization, and Beyond; MIT Press: Cambridge, Mass, 2002.

Smola, A.J.; Scholkopf, B. On a kernel based method for pattern recognition, regression, approximation and operator inversion. *Algorithmica.* 1998, 211-231.

Soulie, M.; Montes, P.; Sivestri, V. Modelling spatial variability of soil parameters." *Canadian Geotechnical Journal.* 1990, 27, 617-630.

Specht, D. F. A general regression neural network. IEEE Trans. *Neural Networks.* 1991, 2, 568-576.

Suykens, J.A.K.; De, B.J.; Lukas, L.; Vandewalle, J. Weighted least squares support vector machines: robustness and sparse approximation. *Neurocomput.* 2002, 48(1–4), 85–105.

Tang, W.H. Probabilistic evaluation of penetration resistance. *Journal of Geotechnical Engineering ASCE.* 1979, 105(GT10), 1173-1191.

Uzielli, M.; Vannucchi, G.; Phoon, K.K. Random filed chracterisation of strees-normalised cone penetration testing parameters. *Geotechnique.* 2005, 55(1), 3-20.

Vanmarcke, E.H. Probabilistic Modeling of soil profiles. *Journal of Geotechnical Engineering ASCE.* 1977, 102(11),1247-1265.

Vanmarcke, E.H. (1983). Random fields: Analysis and synthesis. The MIT Press, Cambridge, Mass.

Vapnik, V.; Lerner, A. Pattern recognition using generalizd portait method. *Automation and Remote Control.* 1963, 24, 774-780.

Vapnik, V.N. Statistical learning theory; Wiley: New York, 1998.

Wahba, G. A comparison of GCV and GML for choosing the smoothing parameters in the generalized spline-smoothing problem. *Ann. Stat.* 1985, 4, 1378-1402.

Wu, T.H.; Wong, K. Probabilistic soil exploration: a case history. *Journal of Geotechnical Engineering ASCE.* 1981, 107(GT12), 1693-1711.

Yaglom, A.M. Theory of stationary random functions. Prentice-Hall Inc.: Englewood Cliffs; NJ, 1962.

In: Machine Tools: Design, Reliability and Safety
Editor: Scott P. Anderson, pp. 187-192
ISBN 978-1-61209-144-0
© 2011 Nova Science Publishers, Inc.

Chapter 7

DOES MINIATURIZATION OF NC MACHINE-TOOLS WORK?

Samir Mekid
KFUPM, Mechanical Engineering Department,
Dhahran, KSA

ABSTRACT

An immediate need was expressed recently and has triggered the process of miniaturization of machines extensively over the last decade. The requirement was about micro-mesoscale components with high to ultra high precision dimensions and surface finish but achieved at low cost according to comparison studies between standard scale machines and desktop machines [1].

The interest is for the development of hybrid subtractive/additive desktop meso/micro NC Machine-tool capable to secure a continuum of manufacturing capabilities and constitute a bridge between well developed technologies at both scales. Although the concept of miniaturization is reasonably justifiable, the challenge is whether in practice current miniaturized machines can achieve the expected requirements?

The next concern is whether the microfabrication techniques at meso-scale level, such as lithography, LIGA and their variations to produce MEMS type components and others are capable to improve and deliver better components in terms of complexity of shapes and dimensional precision.

1. INTRODUCTION

The miniaturization of machine components is currently perceived as a core requirement for the future technological development for meso/microscale features. The mesoscale is defined between microscale and macroscale of the mechanical properties. It is a bridge to Nanotechnology by its natural position within the scale hierarchy. The emerging miniaturised 'high-tech' products are required to have increased functionalities of systems within a size on the order of 1 cm^3. This technology promises enhanced health care and safety, better quality of life, and economic growth by providing micro systems such as bio-medical MEMS, labon-chips in a growing market. It requires mesoscopic parts with complex microscopic features of a few mm length with machining accuracy of less than 1 micrometre and surface integrity as components will require high surface finish, tensile stress and crack free surfaces in order to function reliably. At present, the production of micro-parts is by standard sized machine tools with a modification to the tool size. They consume unnecessarily high energy and occupy a large space and require a large air-conditioned room. Therefore, it is important to consider the development of tailor made micro-machine tools that can save energy, space and resources in controlling the environment e. g. temperature by applying the new concepts of micro-factory and machining with suitable volume-ratio of machine/workpiece discussed in [14]. Traditional fabrication methods (such as the use of focused ion beam, electron beam) of microstructures are often very expensive requiring a vacuum chamber and are slow, and difficult to fabricate complex 3D shapes directly. The required mesoscale parts have delicate features that will require five axis machining and tight tolerances. It is inefficient to produce parts such as moulds and dies with electrical discharge machining or investment casting techniques, as is the current practice. Hence, it becomes essential to develop flexible configurable and cost-effective miniaturized production units. Therefore, accurate micro machining with one or more processes becomes an important requirement in the new concept of the micro-factory. This is an emerging research activity with currently limited knowledge on barriers to improve the accuracy on the existing few examples.

The proposal aims to research generic and creative concepts of machine structures to build a manufacturing intelligent desktop centre capable to host several micromachining processes (hybrid processes) e. g. micromilling, microturning and laser processes individually or in combination and able to deliver high performance micro/meso machining. Laser will be used as a complementary process to micromilling; making holes less than 100 μm, and deburring after micromilling and building up micro-features by additive process on top of the milled component. The milling process, much precise than laser, could also adjust features dimensions built up by laser deposition. The new machine will also host manipulation and inspection as embedded facilities within a versatile platform to ease workpiece handling and measuring. Key characteristics will be initially secured such as high stiffness in machine axes (better than 100 N/μm without the spindle), high stability in positioning and tracking to secure nanometre uncertainty in positioning with superior dexterity of the tool holder with its spindle depending on the process mode. The research will generate design knowledge to permit rapid advances in the design of desktop hardware for meso/micro machining. There is an extreme need to understand fundamental differences in the machine design technique transition from macro to meso/microscale as well as the dominant phenomena affecting the machine performance at small scale. As a consequence, this project will result in a low cost production for on-demand micro-machining of 3D geometric components, allowing for the use of a wide range of materials (metals, ceramics or polymer), an advantage over silicon based micro-fabrication processes. Moreover, it will enable better control of the product capabilities. Low cost and time saving will be achieved by high speed machining due to reduction in distance and mass with an increase of precision by smaller forces and high natural frequencies of the structure. The hybridisation in machine structural concepts reduces machine inertias, thus allowing the control of higher accelerations compared to those permitted by conventional machines. The vibration amplitudes of small machine tools are lower than those of larger machine tools because the inertial forces decrease as the fourth power of the scaling factor and the elastic forces decrease as the second power of the scaling factor. But, these remain significant effects inducing errors at small scale, hence requiring further investigation.

2. INTERNATIONAL CONTEXT OF THE RESEARCH IN THIS FIELD

Non-lithographic (material removal using cutting tools) machining with standard sized machine tools consumes much energy, space and material, while the lithographic method seems to be more advanced but has some limitations in size and dimension (2D, 2.5D). The micro-machining refers to the mechanical material removal using a tool. Most of current micro-machining is provided by a machine having a standard size. A great interest is for portable machines. The PI has had various links and *discussions with European and Japanese micromachine builders*. The Japanese have realized the first portable machining micro-factory [14,15,16] which consists of two or three micro-machines: a lathe, milling and/or press machine with a manipulator. It uses 3 CCD miniature cameras to ease the handling of micro-parts. This micro-factory has produced ball bearings with a diameter of 0.9 mm [17]. In single process machines, Kitahara et. al.[18] have developed a micro-lathe with an X-Y driving unit comprising a slider and a V-shaped guide. A brass rod of 2mm diameter was obtained with Ra=1.5 μm and roundness = 2.5 μm. With the suitability of the inherent excellent characteristics on the micro-machines, the existing machines did reach the required level of accuracy. Mishima [19] stated that design and optimization were not studied enough. The portable prototype exhibited in EUSPEN conference 2004 (Glasgow) by Ito [20] has demonstrated that with an improved model based on the previous lathe [18], the new micro-lathe could achieve on a 10 mm stainless steel an estimated Ra = 60 nm and a circularity of 50 nm. When the small machine tools are made with expensive components and materials it is expected to make savings in a long term run. In the other side, savings in energy and materials could be achieved rapidly with the concept of designing a simple micro-machine, which size is equivalent to the workpiece as proposed by researchers in Mexico [21]. The commercial micro-milling machine proposed recently by Nanowave at EUSPEN 2005 claims 1 μm accuracy made with 3-stacked linear axes but the inspection of the machined parts has revealed a few μm accuracies. For approximately the same size as standard machines, Fanuc has recently commercialised 'Robonano' achieving ultra high precision micro-machining with milling, shaping and grinding and capable of producing mirrors, gratings, lenses...etc. The linear axes have a resolution of 1 nm while rotary axes have $0.0001°$, but the machine is very expensive and available on demand. A number of European machines with standard sizes dedicated to micromachining exist in the UK market such as KERN (CH) (Pyramid) and Kugler (D) (microgantry), although with very good performance, they have difficulties in achieving micrometer *uncertainty* on the workpiece with such a standard machine size.

Some of the existing machines claiming 1μm accuracy still have difficulties in achieving this. It is important to note the lack of such development in the UK. The trend in machining accuracy is to reach sub-micrometer scale in the next decade with full understanding of machine behaviour at this scale.

3. CURRENT LIMITATIONS AND CHALLENGES

Currently, the primary technologies used in industry for miniaturization are the microelectronic fabrication techniques (at meso-scale level), such as lithography, LIGA and their variations to produce MEMS. The principal shortcomings associated with such technologies for MEMS (up to 99% of commercial MEMS production uses LIGA technology) are related to the inability of this technology to produce arbitrary 3D sculptured form features in not only electronics (Silicon-based material) but also in a wide range of metallic and non-metallic materials (stainless steel, brass, gold, titanium, plastic and glass). The new generation of MEMS will have enhanced functions (mechanical, physical and control) and will be strongly based on three-dimensional shapes micromechanical components fully embedded in intelligent electronic devices to create the so called micro-mechatronics devices (MMD). Therefore, a new challenge is to miniaturise mechanical milling and turning machines to reach MEMS scale. Moreover, while conventional microfabrication techniques (e. g. LIGA) can achieve excellent absolute tolerances, relative tolerances are rather poor compared to those achieved by more traditional metal cutting techniques at macro scale level. In traditional macro-machining, relative tolerance of 0.0001% of an accuracy/part-size ratio are becoming standard, whereas in the Electronics/Integrated Circuit industry a 1% relative tolerance is considered to be good. Then, parts with relative tolerance of 0.01% do actually exclude LIGA where absolute size in one or more dimensions is in the micrometer range. Alternative processes like micro electro discharge machining (micro-EDM) can be suitable just for workpieces with 2D symmetry on conductive materials; but leaves heat affected zones and residual tensile stress. The current developments are often driven by the philosophy of using regular size (and high cost) equipment to produce miniature components and devices. This, in turn, has resulted in emphasizing ultra-precision machine tools instead of looking toward the possibility of developing new cost-effective and eco-efficient manufacturing processes to achieve the required levels of precision, accuracy and low-cost productivity in 3D complex shapes component machining. Apart from current research, 'big-size' machines are often used in industry to fabricate micro-parts where cutting forces are in the milli- to a few Newton range. Indeed, hexapod based machining centres have attracted a lot of attention in the machine tool world; they scale exceptionally well for small applications (e. g. micro-manipulators). A small variation in the manufacturing process caused by material or cutting tool characteristics, thermal variations in the machine, vibration and any second order physical phenomenon would have a direct impact on the ability to produce the required features in mass production. These have not been considered in the current machine tools for micro-applications. A sensitivity analysis in the modeling taking into account of those phenomena in the design of next generation micro-machine tools would be another challenge. The effect of the machine performance/behaviour and cutting force on the machined micro part is a challenge to address. Handling of micro parts, if not robust, would have an impact on the repeatability of a process that has a desired tolerance of less than a micron. The usual gripping/holding technique would not apply to micro-parts. Adapting metrology to inspect workpieces and systematic machine accuracy maintenance is another challenge. In addition, machine tools are only capable of taking materials off (i. e. subtractive machining). The combination of laser micro-deposition and micro-CNC milling/turning is a new concept that could provide the possibility of making a component with different materials at different locations and reducing the total machining time.

REFERENCES

[1] S. Mekid, Design Strategy for Precision Engineering: Second Order Phenomena, *J. Engineering Design*, vol.16. no: 1, 2005. pp 63-74

[2] S. Mekid, A. Gordon and P. Nicholson, Challenges and Rationale in the Design of a Miniaturised Machine Tool, International MATADOR, Conference, 2004, UMIST Manchester, UK.

[3] S. Mekid, High Precision Linear Slide. Part_1: Design and Construction, *Int. Journal of Machine Tools and Manufacture* 40(7) (2000) pp.1039-1050.

[4] S. Mekid O. Olejniczak High Precision Linear Slide. Part_2: Control and Measurements. *Int. Journal of Machine Tools and Manufacture* 40(7) (2000) pp.1051-1064.

[5] S. Mekid and M. Bonis, Conceptual Design and study of high precision translational stages: Application to an optical delay line" *J. American Society for Precision Engineering*, Vol. 21 No 1, July 1997. pp. 29-35.

[6] A. Khalid S. Mekid, Design of precision desktop machine tools for meso-machining, Proceedings of the 2nd Virtual International Conference on Intelligent Production Machines and Systems, Elsevier (Oxford) 2006.

[7] A. Khalid S. Mekid, Design and Optimization of a 3-axis Micro-Milling Machine, 6th Int. Conf. European Society for Precision Engineering and Nanotechnology, Baden, Austria May 2006.

[8] S. Mekid, A. Khalid, Robust Design with Error Optimization Analysis of CNC Micro Milling Machine, 5th CIRP International Seminar on Intelligent Computation in Manufacturing Engineering - CIRP ICME '06, 25-28 July 2006, Ischia (Naples), Italy.

[9] A. Khalid, S. Mekid, Analysis of Jacobian inversion in parallel kinematic systems, 35th International MATADOR Conference 2007, Taiwan.

[10] Alexander H. Slocum, *Precision Machine Design*, Prentice Hall, 1992.

[11] Alec P. Robertson and Alexander H. Slocum Measurement and characterization of precision spherical joints, *Precision Engineering*, Volume 30, Issue 1, January 2006, Pages 1-12

[12] Eberhard Bamberg, Christian P. Grippo, Panitarn Wanakamol, Alexander H. Slocum, Mary C. Boyce and Edwin L. Thomas A tensile test device for in situ atomic force microscope mechanical testing, *Precision Engineering*, Volume 30, Issue 1, January 2006, Pages 71-84.

[13] Jean-Sébastien Plante, John Vogan, Tarek El-Aguizy and Alexander H. Slocum, A design model for circular porous air bearings using the 1D generalized flow method, *Precision Engineering*, Volume 29, Issue 3, July 2005, Pages 336-346.

[14] Okazaki Yuichi et al., 2004, Microfactory—Concept, History, and Developments, *Journal of Manufacturing Science and Engineering*, Vol. 126, pp837-844.

[15] Oyama Naotaki et al.,2000, Desktop machining micro-factory, Proc. 2nd Int. workshop on micro-factories, Switzerland,pp14-17.

[16] M. Tanaka,, Development of desoktop machining microfactory, Riken Review No 34. (April 2001).

[17] Maekawa H.,Komoriya K., 2001, Development of a micro-transfer arm for a micro-factory, *Proc. of IEEE international conference on robotic sand automation*, Seoul, Korea.

[18] T. Kitahara, Y. Ishikawa, K. Terada, N. Nakajima and K. Furuta, 1996, Development of Micro-lathe, *Journal of Mechanical Engineering Laboratory*, Vol. 50, No. 5, pp. 117-123.

[19] Mishima N., Ashida K., Tanikawa T., Maekawa H., Development of desktop machining microfactory, 2000, Japan-USA Flexible Automation Conference, Michigan.

[20] S. Ito et.al , precision turning on a desk- micro turning system, EUSPEN Conference Glasgow, UK May-June 2004.

[21] E.Kussul et al., 2002, development of micromachine tool prototypes for microfactories, *J. Micromech. and Microengineering*, 12.

[22] D. Dornfield, S. Min and Y. Takeuchi, Recent advances in mechanical micromachining, *Annals of the CIRP*, Vol.55/2, 2006, pp. 745-768.

[23] Mekid S, Ryu H.S, Rapid vision-based dimensional precision inspection of mesoscale artefacts. Proceedings of the Institution of Mechanical Engineers - Part B: Journal *of Engineering Manufacture.* Vol. 221. 2007.

Chapter 8

COMPUTER-CONTROLLED MACHINE TOOL WITH AUTOMATIC TRUING FUNCTION OF WOOD-STICK TOOL

Fusaomi Nagata[1], *Takanori Mizobuchi*[1], *Keigo Watanabe*[2],
Tetsuo Hase[3], *Zenku Haga*[3] *and Masaaki Omoto*[3] *

[1]Tokyo University of Science, Yamaguchi, Japan
[2]Okayama University, Japan
[3]R & D Center, Meiho Co. Ltd., Japan

Abstract

In this chapter, a computer-controlled machine tool with an automatic truing function of a wood-stick tool is described for the long-time lapping process of an LED lens cavity. We have experimentally found that a thin wood stick with a ball-end tip is very suitable for the lapping tool of an LED lens cavity. When the lapping is conducted, a special oil including diamond lapping paste, whose grain size is about 3 μm, is poured into each concave area on the cavity. The wood material tends to compatibly fit both the metallic material of cavity and the diamond lapping paste due to the soft characteristics of wood compared with other conventional metallic abrasive tools. The serious problem in using a wood-stick tool is the abrasion of the tool tip. For example, an actual LED lens cavity has 180 concave areas so that the cavity can form small plastic LED lenses with mass production at a time. However, after about five concave areas are finished, the tip of the wood-stick tool is deformed from the initial ball shape as a result of the abrasion. Therefore, in order to realize a complete automatic finishing system, some truing function must be developed to systematically cope with the undesirable tool abrasion. The truing of a wood-stick tool means the reshaping to the initial contour, i.e., ball-end tip. In the chapter, a novel and simple automatic truing method by using cutter location data is proposed and its effectiveness and validity are evaluated through an experiment. The cutter location data are called the CL data, which can be produced from the main-processor of 3D CAD/CAM widely used in industrial fields. The proposed machine tool can easily carry out the truing of a wood-stick tool based on the generally-known CL data, which is another important feature of our proposed machine tool.

*E-mail address: http://www.meiho-j.co.jp/

1. Introduction

It is not easy to automate the lapping process of a metallic LED lens cavity because the cavity itself is not axis-symmetric and a lot of concave areas on the cavity to be finished are very small, e.g., each diameter is 4 mm. Tsai et al. developed a novel mold polishing robot [1] and its path planning technique [2], however, the applicability to the lapping process of an axis-asymmetric LED lens cavity was not described. Actually, we could not find other previous literature concerning the finishing of an axis-asymmetric LED lens cavity. That is the reason why such an LED cavity is being dexterously finished by the hand lapping of a skilled worker in the related industrial field. However, the long-time hand lapping cannot be successfully carried out even for the skilled worker, consequently, defective products caused by the dispersion of quality tend to appear.

To finish a concave area on an LED cavity, a computer-controlled machine tool has been already proposed [3]. The machine tool is comprised of three single-axis devices with a high position resolution of 1 μm. A thin wood-stick tool is attached to the tip of the z-axis. A compact force sensor with three degree-of-freedoms is built in the z-axis. The machine tool has two functional applications. One is the 3D machining function, which is easily realized by using the position control mode. The 3D machining function can be applied to the truing of a wood-stick tool. The other is the profiling function, which is performed due to the hybrid position/force control mode along a spiral path. The profiling function can be applied to the lapping of concave areas on a metallic LED lens cavity.

We have experimentally found that a thin wood stick with a ball-end tip is very suitable for the lapping tool of an LED lens cavity. When the lapping is conducted, a special oil including diamond lapping paste, whose grain size is about 3 μm, is poured into each concave area on the cavity. The wood material tends to compatibly fit both the metallic material of the LED lens cavity and the diamond lapping paste due to the soft characteristics of wood compared with other conventional metallic abrasive tools. The serious problem in using a wood-stick tool is the abrasion of the tool tip. For example, an actual LED lens cavity has 180 concave areas so that the cavity can form small plastic LED lenses with mass production at a time. However, after about five concave areas are finished, the tool tip of the wood-stick is deformed from the initial ball shape as a result of the abrasion. Therefore, in order to realize a complete automatic finishing system, some truing function must be developed to systematically cope with the undesirable tool abrasion. The truing of a wood-stick tool means the reshaping to the original shape, i.e., ball-end tip.

In the chapter, a novel and simple automatic truing method by using cutter location data is proposed for the long-time lapping process of an LED lens cavity and its effectiveness and validity are evaluated through an experiment. The cutter location data are called the CL data, which can be produced from the main-processor of 3D CAD/CAM widely used in industrial fields. The proposed machine tool can easily carry out the truing of a wood-stick tool based on the generally-known CL data, which is another important feature of our proposed machine tool. The chapter is organized as follows. In section 2, a computer-controlled desktop machine tool is briefly introduced. In section 3, a transformation technique from fine velocity to pulse is proposed for the pulse-based servo controller. In section 4, the control system of the machine tool is described in detail, which has multi-control modes composed of the position control with/without dealing with a machining force and the profiling control along a curved surface. The structure is very simple for the provided functions. For example, 3D machining function is easily realized by using the position control mode. Also, compliant motion to a workpiece is performed due to the profiling control mode along a desired trajectory, which will be able to be applied to the lapping process of a metallic LED lens cavity. In section 5, a finishing experiment is conducted to show the effectiveness of the machine tool. And, an automatic truing function for a thin wood-stick tool is further proposed to achieve the long-time finishing process. The idea and the detailed software flowchart are described. Finally, the conclusions obtained are summarized in section 6.

2. Computer-Controlled Machine Tool

Figure 1. Computer-controlled machine tool with multi-application modes.

Figure 1 shows the developed desktop machine tool consisting of three single-axis devices with position resolution of 1 μm. The size is 850 × 645 × 700 mm. The single-axis device is a position control robot ISPA with high-precision resolution provided by IAI Corp., which is comprised of a base, linear guide, ball-screw and AC servo motor. The effective strokes in x-, y- and z-directions are 400, 300 and 100 mm, respectively. The tool axis is designed to be parallel to z-axis of the machine tool. A wood-stick tool is fixed to the tip through a 3-DOF compact force sensor. To regulate the rotation, a servo motor is located parallel to the tool axis. Figure 2 shows the large scale photo of the tip of the z-axis rotated by the servo motor. The position resolution and force resolution, and effective stiffness at the tool tip were examined through a simple contact experiment, so that the force resolution of 0.178 N was obtained due to the position resolution of 1 μm. Therefore, the effective stiffness can be estimated as 178 N/mm.

Figure 2. Tool head of the proposed machine tool, where a ball-end mill or a thin wood-stick tool is mounted in the attachment.

3. Transformation from Fine Velocity to Pulse

From the viewpoints of programming interface, torque command, position command and pulse command are available for the control of servo motors. Recently, a position-based servo controller in Cartesian coordinate system is provided for articulated industrial robots with open architecture, such as Kawasaki JS10, FS20, Mitsubishi PA10 and Motoman UP6. The manipulated values of position less than the repetitive position resolution guaranteed, e.g., 0.1 mm, are valid and can be given to the servo controller with "float" type or "double" type data in C programming language. On the other hand, the pulse-based servo controller is widely spread to the mechatronics field because of its performance and easiness of the treatment, and is also employed in the servo motor to control a single axis used in the proposed machine tool. The pulse-based servo controller transforms the pulse command into the motor driving torque.

However, when trying to realize a precise position control system by using the pulse-based servo controller, the relationship between one pulse and the corresponding minimum resolution should be noted. The position control of a single axis of the machine tool is performed by the number of pulses, and the relative position per one sampling time can be regarded as the velocity. For example, assume that the minimum resolution in position is 1 μm for a single axis of the machine tool, i.e., the machine tool can move 1 μm by one pulse. In this case, the single-axis does not move, even if the manipulated value of 0.1 μm is continuously given to the pulse-based servo controller every sampling time of 1 msec in order to achieve the velocity control of 100 μm/s. This is attributed to the fact that the pulse-servo controller generally ignores the fraction (i.e., values below decimal point) less than one pulse.

Figure 3 shows an example of the normal velocity $v_{normal}(k)$ that was generated by the impedance model following force controller, according to the force error $E_f(k)$ in a profiling control experiment. The values in Fig. 3 mean relative position commands [μm] per one sampling time of 1 msec. Observe that almost values are plotted within the range of

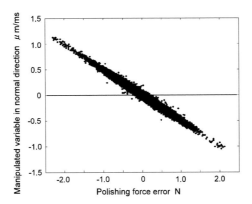

Figure 3. Relation between the polishing force error $E_f(k) = F_d - \|F(k)\|$ and manipulated value $v_{normal}(k)$ calculated from the impedance model force controller given by Eq. (8).

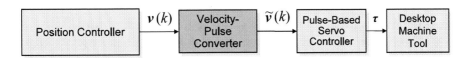

Figure 4. Block diagram of the proposed fine velocity-pulse converter.

± 1 μm. It should be noted, however, that the values less than 1 μm can not be treated with the pulse-based servo controller. To overcome this problem, a transformation technique called the velocity-pulse conversion is proposed here. The velocity-pulse converter transforms the fine manipulated value in velocity into the pulse command given by an integer, which can be outputted to the pulse-based servo controller. The velocity-pulse converter for the pulse-based servo controller is located after the position controller as shown in Fig. 4. Every sampling time period, to obtain the output $\tilde{v}_x(k)$ of the velocity-pulse converter in x-direction, calculate the summation of fractions, $S(k)$, such as

$$S(k) = \sum_{n=k_i}^{k} v_x(n) \qquad (1)$$

where $v_x(n)$ denotes the fraction value at time n and k_i is the latest discrete time when the velocity-pulse converter generated non-zero value. If the absolute value of $S(k)$ is greater than one, then the velocity-pulse converter generates the following pulse:

$$\tilde{v}_x(k) = \begin{cases} \text{Int}(S(k)) & \text{if } |S(k)| \geq 1 \,\mu\text{m} \\ 0 & \text{otherwise} \end{cases} \qquad (2)$$

where Int($\cdot$) is the function that extracts only the part of an integer with sign from the value of $(\cdot)$. Furthermore, the fraction value should be calculated, for the next step, as

$$v_x(k+1) = \begin{cases} S(k) - \tilde{v}_x(k) & \text{if } |S(k)| \geq 1\,\mu\text{m} \\ S(k) & \text{otherwise} \end{cases} \quad (3)$$

where update $k_{i\_new} = k$ as a new nonzero output time when satisfying $|S(k)| \geq 1\,\mu$m. Similarly, $\tilde{v}_y(k)$ and $\tilde{v}_z(k)$ in y- and z-directions are obtained. Thus, by applying the proposed velocity-pulse converter, the velocity vector $v(k)$ in Cartesian space can be transformed into the pulse command vector $\tilde{v}(k) = [\tilde{v}_x(k)\ \tilde{v}_y(k)\ \tilde{v}_z(k)]^T$ which can be given to the pulse-based servo controller.

4. Control System for Multi-manufacturing Mode

In this section, three types of machining modes are explained. Each machining mode allows the machine tool to perform different types of applications. Figure 6 shows examples of the orthodox 3D machining mode and the profiling control mode.

4.1. 3D Machining Mode

In 3D machining mode, a position feedforward control is used. Position components in multi-axis CL data as shown in Fig. 5 are used for the desired trajectory of a router bit as shown in Fig. 7. Figure 8 shows the block diagram of the controller. Now, the position vector of the tool tip at the discrete time k is given by $x(k) \in \Re^3$. The i-th step of CL data is $x_{cl}(i) = [p(i)\ o(i)]^T$. It is assumed that $x(k) \in [x_{cl}(i), x_{cl}(i+1)]$, where $p(i) \in \Re^3$ and $o(i) \in \Re^3$ are the position vector and normalized direction vector, respectively. The direction is normal to the surface of a workpiece. When the tangent vector is obtained from $t(i) = x_{cl}(i+1) - x_{cl}(i)$, then the tangent velocity $v_t(k) \in \Re^3$ is written by

$$v_t(k) = v_{ts}\frac{t(i)}{\|t(i)\|} \quad (4)$$

where v_{ts} denotes the scalar of feed rate [mm/s]. In machining, a constant value is set to v_{ts}. Figure 9 shows the update timing in case that the CL data are calculated with a linear approximation method. The update timing means when the next step in CL data is set. If the

```
Cutter Location data
GOTO/-0.0003,-129.7928,9.7667, 0.0000086,0.7684358,0.6399269
GOTO/0.3104,-129.1995,9.0583, -0.0076906,0.7660909,0.6426862
GOTO/-0.3085,-129.2005,9.0594, 0.0076439,0.7660970,0.6426795
GOTO/-0.0003,-129.7928,9.7667, 0.0000086,0.7684358,0.6399269
```

Figure 5. Example of CL data consisting of position and direction components.

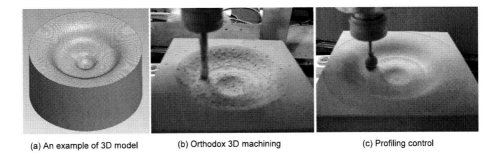

(a) An example of 3D model (b) Orthodox 3D machining (c) Profiling control

Figure 6. Examples of 3D machining control and profiling control by using the machine tool.

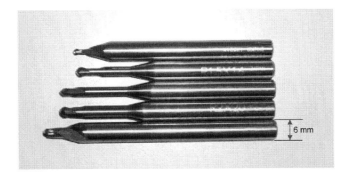

Figure 7. Router bits used in 3D machining mode.

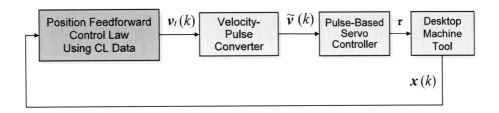

Figure 8. Block diagram of the position feedforward controller using the position components in CL data.

linear approximation is applied, CL data forming a curved line consist of these continuous minute lines such as $x_{cl}(i) - x_{cl}(i-1)$ and $x_{cl}(i+1) - x_{cl}(i)$. $x_d(k)$ is the desired position at the discrete time k calculated along CL data. $\|x_d(k) - x_d(k-1)\|$ is determined by a given feed rate. In order to realize precise position control along CL data, in case of the example as shown in Fig. 9, $x_{cl}(i)$ is set to $x_d(k+1)$ just before the feed direction changes. Accordingly, the fraction $\|x_{cl}(i) - x_d(k)\|$ is used as the manipulated value of this case.

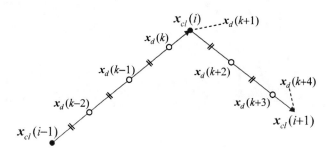

Figure 9. Update timing of CL data from $x_{cl}(i-1)$ to $x_{cl}(i)$.

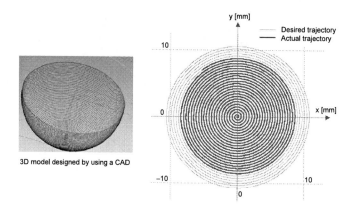

Figure 10. Position control result viewed in x-y plane along a spiral path, in which the controller shown in Fig. 8 is used.

In 3D machining mode, a model designed by 3D CAD can be machined by moving a router bit such as a ball-end mill along CL data. Figure 10 shows an example concerning a desired trajectory and an actual trajectory viewed in x-y plane, in which the position of the router bit is controlled along a desired trajectory (CL data) generated with a spiral path of 0.4 mm pitch. The actual trajectory means the controlled one. It is observed that the router bit can satisfactorily follow the desired trajectory. The 3D machining mode will be able to be applied to the automatic truing function of a wood-stick tool with a ball-end shape as shown in Fig. 11.

4.2. 3D Machining Mode Handling Machining Load

When the proposed 3D machining mode shown in Fig. 8 is used, the machining load [N] against the router bit used varies every moment depending on the change of the cutting

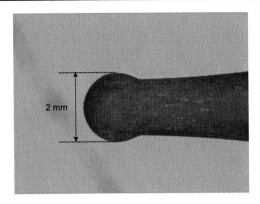

Figure 11. Wood-stick tool used in the lapping process of a metallic cavity, whose tip has a small ball-end shape of 2 mm.

amount according to each designed model. Because of this, not only uncomfortable noise and vibration but also fatal damage of the router bit or the workpiece tend to occur. In this subsection, to overcome the problems, another 3D machining mode handling the machining load is proposed. As a detecting method of the machining load, the load torque [N·m] obtained from the voltage change of a tool spindle is used in general. On the contrary, a compact 3-DOF force sensor is employed in the proposed machine tool because of the connection abilities with a personal computer and a main-head located in z-axis. Therefore, the machining load described in the chapter is generated by the translational motion of a router bit, and means each component in x-, y- or z-direction measured by the force sensor. The machining force [N] is further defined as the resultant force of x-, y- and z-directional force sensor measurements in sensor coordinate system. The 3D machining mode handling the machining loads enables a novel manufacturing method keeping the machining force to be a desired value. As another effectiveness, it is expected that undesirable deformation may be suppressed when an elastic material such as a rubber is machined.

First of all, let us consider the following velocity-based control law for 3D machining using the machining force. As can be seen, the controller can be derived intuitively.

$$\tilde{\boldsymbol{v}}_t(k) = \boldsymbol{v}_t(k) - \boldsymbol{K}_p \left(\|\boldsymbol{F}(k)\| - F_d\right) \frac{\boldsymbol{t}(i)}{\|\boldsymbol{t}(i)\|} - \boldsymbol{K}_i \sum_{n=1}^{k} \left(\|\boldsymbol{F}(n)\| - F_d\right) \frac{\boldsymbol{t}(i)}{\|\boldsymbol{t}(i)\|} \quad (5)$$

where $\boldsymbol{K}_p \in \Re^{3\times 3}$ and $\boldsymbol{K}_i \in \Re^{3\times 3}$ are the proportional and integral gains, which are set to positive-definite diagonal matrices, respectively. $\boldsymbol{F}(k) \in \Re^3$ is the force vector measured by the force sensor. $\|\boldsymbol{F}(k)\|$ is the machining force. Also, F_d is the desired machining force given by an operator. Figure 12 shows the block diagram of the position feed forward controller using the position components in CL data and machining force. According to the machining force, this controller compensates the position of a router bit in the feed direction written by $\boldsymbol{t}(i)$. Hence, undesirable shape error between a model designed by 3D CAD and its machined shape is foreseen since the compensation disturbs $\boldsymbol{v}_t(k)$. Actually, we conducted a machining test using a material of chemical wood, so that a shape error was confirmed clearly.

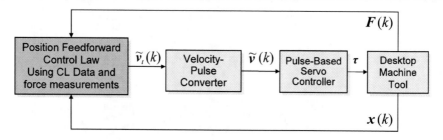

Figure 12. Block diagram of the position feed forward controller using the position components in CL data and the machining force.

In order to overcome the problem, an improved control law is further proposed. The controller regulates the scalar v_{ts} in Eq. (4) as given by

$$v_{ts} = \begin{cases} v_{ts} \dfrac{F_d}{\|\boldsymbol{F}\|} K_{cl} & \text{if } \|\boldsymbol{F}\| > F_d \\ v_{ts}(1 + (F_d - \|\boldsymbol{F}\|)K_{ch}) & \text{if } \|\boldsymbol{F}\| < F_d \end{cases} \quad (6)$$

where K_{cl} and K_{ch} are the gains to tune the decay rate and the amplification rate, respectively. Due to the proposed method regulating the machining force, a skillful 3D machining suppressing undesirable deformations can be realized even in the case of a workpiece as a thin wood-stick tool which is easy to bend.

4.3. 3D Profiling Control Mode

In this subsection, 3D profiling control mode is introduced. The profiling controller allows a tool to move along curved surface of a workpiece keeping the contact force to be a desired value, so that the application for polishing task will be able to be realized. At first, a 3D profiling control mode is designed as shown in Fig. 13 by adding a force feedback loop to the position feedforward controller shown in Fig. 8. $\boldsymbol{v}_n(k)$ is the manipulated variable for the force feedback control, which is given by

$$\boldsymbol{v}_n(k) = v_{normal}(k)\boldsymbol{o}(k) \quad (7)$$

where $\boldsymbol{o}(k)$ is the normalized vector in normal direction calculated from orientation components in CL data. $v_{normal}(k)$ representing a scalar in normal direction is the output from the impedance model force control [4] given by

$$v_{normal}(k) = v_{normal}(k-1)\, e^{-\frac{B_d}{M_d}\Delta t} + \left(e^{-\frac{B_d}{M_d}\Delta t} - 1\right)\frac{K_f}{B_d}(F_d - \|\boldsymbol{F}(k)\|) \quad (8)$$

where K_f is the force feedback gain, and impedance parameters M_d and B_d are the desired inertia and damping coefficients, respectively. F_d is the desired polishing force; Δt is the sampling time. Although the standard Windows timer was set to 10 ms, the actual rate measured by a data logger was about 15 ms. So, we used the Windows multimedia timer to realize 1 ms sampling rate.

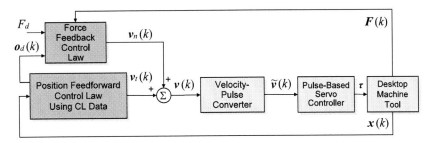

Figure 13. Block diagram of the position feed forward controller with a force closed loop.

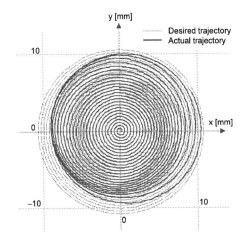

Figure 14. Position control result viewed in x-y plane in case that the controller shown in Fig. 13 is used for a profiling control.

A profiling control experiment was conducted by using the controller shown in Fig. 13. In this case, a spiral path leading from the center to the outside is used for the desired trajectory. Figure 14 shows the control result of the tool tip position viewed in x-y plane. It is observed that, since the manipulated value $v_n(k)$ is generated from the force feedback loop in profiling control mode, the feedforward position controller referring the desired trajectory made of CL data cannot achieve a good result. Of course, if the controller is tried to be applied to some automatic polishing system, then such a deviation from a desired trajectory will bring out undesirable ununiformity on the workpiece's surface.

4.4. 3D Profiling Control Mode Considering the Read Timing of CL Data

In the profiling control mode to be applied to a polishing robot and so on, the force control system must be overridden. Therefore, in the previous subsection, the tool position was moved by the feedforward controller so as not to interfere with the force feedback loop. However, this causes a problem that the manipulated value $v_n(k)$ from the force feedback

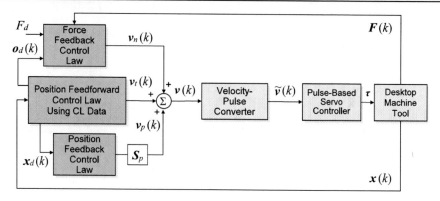

Figure 15. Block diagram of the profiling control system with two closed loops for force and position.

loop gives bad influence to the position feedforward system. To cope with the problem, how to update the step $x_{cl}(i)$ in CL data, i.e., the read timing of the next step in CL data, is considered.

Here, we pay attention to the stick-slip motion which has been known as an effective method to improve the polishing efficiency [5]. If it is assumed that the small stick-slip motion yielding in the orthogonal direction to tool's moving one can increase the polishing performance, it may be not required that such a strict position control along CL data as shown in Fig. 9. In other words, small vibrational-like deviations against a desired trajectory can be regarded as an expectable small stick-slip motion.

In Fig. 13, the tool position is controlled by $v(k)$ which is the sum of the manipulated variable $v_n(k)$ from the force feedback loop and $v_t(k)$ from the position feedforward loop. Accordingly, a simple method on how to determine the read timing can be proposed. The method is that when the following condition is satisfied, the next step $x_{cl}(i+1)$ in CL data is immediately read in.

$$\|\sum_{n=k_i}^{k} v(n)\| \geq \|t(i)\| \qquad (9)$$

where k_i denotes the time when $x_{cl}(i)$ was read in, i.e., last update was conducted. Due to the above improvement, even if the tool moves with some deviation from a desired trajectory, the update from $x_{cl}(i)$ to $x_{cl}(i+1)$ can be conducted just when the tool passes around $x_{cl}(i)$ shown in Fig. 9.

A position feedback loop written with a simple PI control law is further added as shown in Fig. 15 so that the tool can spirally rise up accurately keeping the position of z-directional components in CL data. Figure 16 shows an example of a controlled trajectory viewed in x-y plane, in which the controller shown in Fig. 15 is used. It is observed that the profiling control is successfully achieved clearly following the desired spiral path.

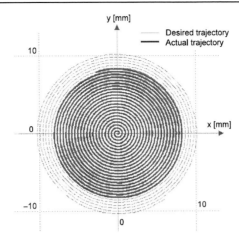

Figure 16. Controlled trajectory viewed in x-y plane, in which the controller shown in Fig. 15 is used.

5. Automatic Truing

5.1. Finishing Experiment

Next, the effectiveness of the machine tool with the controller shown in Fig. 15 is examined through an actual lapping test of an LED lens cavity. Figure 17 shows an example of skilled worker's hand lapping scene, where an LED lens cavity with multiple concave areas is being finished. The lapping process has been being required to be automated, however, its complete achievement is not easy even at the present stage. Figure 18 shows the lapping scene by using the machine tool with the controller shown in Fig. 15. In this case, the polishing force acting between the tool and the LED lens cavity was controlled as shown in Fig. 19, in which the desired value was set to 20 N. The high frequency vibrations are caused by the tool rotation of 400 rpm. Figure 20 shows the surfaces before and after the lapping process. It is observed that the undesirable cusps can be almost removed uniformly. It has been confirmed from the result that the proposed machine tool has an effectiveness to achieve a higher quality surface.

5.2. Automatic Truing Function for a Thin Wood-Stick Tool

In the previous subsection, a finishing process of an LED lens cavity is introduced as one of applications of the proposed machine tool. In order to apply the machine tool to an actual industrial manufacturing line, the problem on the abrasion of a wood stick tool should be overcome. Finally, an interesting and important peripheral technique is described briefly. Here, an automatic truing function is proposed to be able to continuously finish plural concave areas machined on an LED lens cavity. Although the tool tip has a ball-end shape with a radius of R [mm] at the start time of the finishing, its ball shape is gradually deformed by the contour abrasion with the passage of the finishing time. Therefore, to keep the initial performance of the tool, the truing of the tool tip plays a vital role. Figure 21 shows the

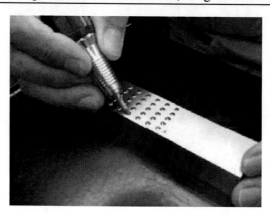

Figure 17. Skilled worker's hand lapping scene of an LED lens cavity with plural concave areas to be finished.

Figure 18. Automated lapping scene of a concave area on an LED lens cavity.

proposed simple scheme for an automatic wood-stick tool truing. A tool bit for automatic truing is precisely fixed on the table of the machine tool, so that the contour of the wood-stick tool can be easily reformed to the initial shape, i.e., the radius of R, by moving the rotating abraded tool along the trajectory from point (A) to point (B). Note that the desired trajectory written with a dotted line, where the tool with a radius of R is reversed in top and bottom, can be easily drawn by using 3D CAD and the CL data are generated by using the main-processor of the CAM.

Figure 22 shows the truing scene of a wood-stick tool, in which its desired shape of contour is illustrated. The CL data used is generated within the area from point O to point P. As can be seen, the tool is being reshaped like the CAD model. The diameter of the wood-stick tool is 3 mm, so that the stiffness in the side directions is lower than the one in the vertical direction. Hence, if the truing is not conducted with a small machining force, then undesirable tool flexure will appear. The machining method used is given by Eqs. (4) and (6); the desired machining force is set to 5 N. It has been confirmed from the experiment that the automatic truing function is successfully realized by using the proposed method.

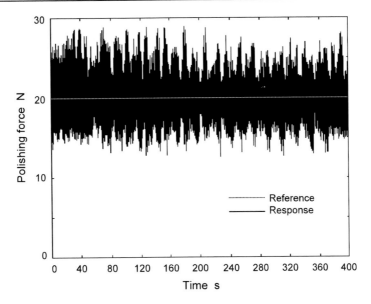

Figure 19. Control result of polishing force acting between the tool tip and the LED lens cavity.

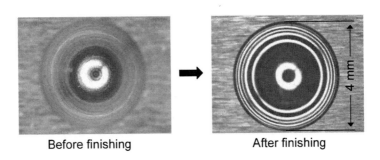

Figure 20. Finished surface of a concave area on an LED lens cavity, in which the controller shown in Fig. 15 is used.

Figure 23 shows the software flowchart of the machine tool in running. First of all, if all concave areas, e.g. 180, are all finished, then the task is completed. Otherwise, the ball-end tip of the wood-stick tool puts the diamond paste on, and goes to the slightly above position of the next concave area. Then, the tool tip approaches to the concave area with a low speed of 1 mm/s. After detecting a contact with the center of the concave area, a lapping action based on the control scheme shown in Fig. 15 starts immediately. Next, the truing of the wood-stick tool is conducted by using the method as introduced in this section, and the completion of the task is checked at the top of the flowchart.

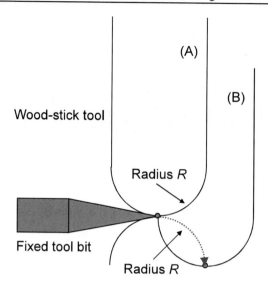

Figure 21. Simple idea for realizing an automatic tool truing.

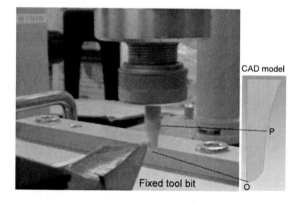

Figure 22. Truing scene of a wood-stick tool and its desired shape of contour.

6. Conclusion

In this chapter, a computer-controlled machine tool with multi-control modes has been presented. The multi-control modes include position control with/without handling the machining force, profiling control along a curved surface, and so on. The machine tool consists of three single-axis devices with a high position resolution of 1 μm. A ball-end mill called a router bit or a thin wood-stick tool can be attached to the tip of the z-axis. The proposed machine tool has realized compliant and dexterous motion required for the skilled tasks such as force-controlled 3D machining and profiling control along a curved surface. Especially, it has been confirmed that the profiling control mode can be applied to the lapping process of a metallic LED lens cavity.

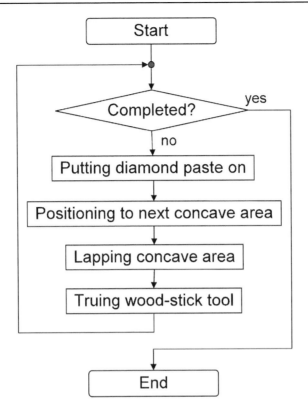

Figure 23. Software flowchart of the machine tool for realizing a long-time automatic lapping process.

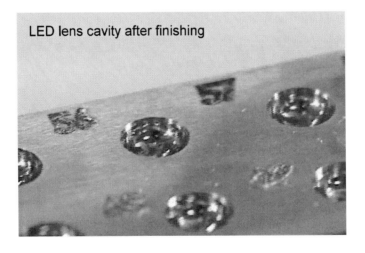

Figure 24. An LED lens cavity finished by the proposed machine tool.

Further, a novel and simple automatic truing method by using cutter location data is proposed and its effectiveness and validity are evaluated through an experiment. The cutter location data is called the CL data, which can be produced from the main-processor of 3D CAD/CAM widely used in industrial fields. The proposed machine tool can easily carry out the truing of a wood-stick tool based on the generally-known CL data, which is another important feature of our proposed machine tool. By using the proposed machine tool introduced in this chapter, such a high quality surface as shown in Fig. 24 can be automatically obtained.

Finally, we hope that the proposed system will be also useful for not only engineers in related industrial fields but also students pursuing education and research concerning machine tool, 3D machining, advanced control system, CAD/CAM, automation and so on.

Acknowledgments

This work was supported by Grant-in-Aid for Scientific Research (C) (20560248) of Japan.

References

[1] Tsai, M.J., Chang, J.L. and Haung, J.F., *Procs. of 2003 IEEE International Conference on Robotics & Automation*, 2003, pp. 3517–3522, Development of an Automatic Mold Polishing System.

[2] Tsai, M.J., Fang, J.J. and Chang, J.L., *International Journal of Robotics & Automation*, 2004, vol. 19, no. 2, pp. 81–89, Robotic Path Planning for an Automatic Mold Polishing System.

[3] Nagata, F., Hase, T., Haga, Z., Omoto, M. and Watanabe, K., *Artificial Life and Robotics*, 2009, vol. 13, no. 2, pp. 423–427, Intelligent Desktop NC Machine Tool with Compliant Motion Capability.

[4] Nagata, F., Hase, T., Haga, Z., Omoto, M. and Watanabe, K., *Mechatronics*, 2007, vol. 17, nos. 4–5, pp. 207–216, CAD/CAM-Based Position/Force Controller for a Mold Polishing Robot.

[5] Nagata, F., Mizobuchi, T., Tani, S., Hase, T., Haga, Z., Watanabe, K., Habib, M.K. and Kiguchi, K., *Procs. of 2010 IEEE International Conference on Robotics and Automation* , 2010, pp. 2095–2100, Desktop Orthogonal-Type Robot with Abilities of Compliant Motion and Stick-Slip Motion for Lapping of LED Lens Molds.

In: Machine Tools: Design, Reliability and Safety
Editor: Scott P. Anderson, pp. 211-231
ISBN: 978-1-61209-144-0
© 2011 Nova Science Publishers, Inc.

Chapter 9

FAULT MONITORING AND CONTROL OF MECHANICAL SYSTEMS

S. N. Huang and K. K. Tan
National University of Singapore
Singapore 117576

Abstract

In machine tools, the designed systems include many components, such as sensors, actuators, joints and motors. It is required to all these components should work properly to ensure safety. In this chapter, we will study fault monitoring and control scheme in the machine system, detecting machines whenever a failure occurs and accommodating the failures as soon as possible. In this scheme, the monitoring algorithm is first designed based on the system model. Under such circumstances, when a failure occurs, the fault-tolerant control scheme is activated to compensate the effects of the fault function. Case study is given to illustrate the performance of the designed monitoring and controller in a real-time machine.

1. Introduction

Industrial machine systems include many components, such as sensors, actuators and motors. These components are required to function according to some specifications and requirements in order for the overall system to operate precisely and reliably. However, due to prolonged operations or harsh operating environment, the performance of these devices may degrade to an unacceptable level. At this stage, more regular faults occur leading ultimately to a total collapse in function. The costs to be incurred in bringing back the system to work are far higher than the direct costs due to rectification and parts replacement, including production loss which can last over a long duration.

Various approaches to fault detection have been reported over the last two decades. It has been shown that the use of adequate models can allow early fault detection with normal measurable variables [1]. In [2], an expert system model is developed for fault detection. In [3], a parameter estimation based on block-pulse function series is used to detect and diagnose faults in permanent magnet DC motor. In [4], a chip thickness and cutting force model is built for predicting process faults. In [5], a robust fusion approach based on fuzzy logic is developed for reliable machinery health assessment. In [9], we develop a linear state observer for detecting the cutting tool wear. In [6], an adaptive observer technique is proposed for a fault diagnosis of actuators. In [7], a dynamical model is presented to detect incipient faults. In [29], a model following method is used for reconfiguration of flight control in the face of structural damage or system faults. In [8], a technique to improve the fault detection is presented by using the classical multiple signal classification (MUSIC) method. However, the model-based fault detection schemes depend on the assumption that a mathematical characterization of the robotic system is available. In practice, this may not hold since it is difficult to obtain an accurate model. Recently, nonlinear approximation approaches to nonlinear fault detection problem have been developed [30, 10, 11, 12, 13, 14, 15, 16, 17]. Furthermore, in many applications, it is important not only to detect but also to accommodate a failure as quickly as possible. Visinsky *et al* [2] propose an expert system for fault tolerant control (FTC). In [19], adaptive methods for accommodating actuator failures are studied. In [23], a fault diagnosis and tolerant control approach is presented which is based on a simple first order system. In [20], stable adaptive controllers are applied to achieve fault-tolerant engine control. Most of these studies are focused on single-input-single-output (SISO) systems [23, 19, 20]. The FTC problem that arises in multiple-input-multiple-output (MIMO) systems introduces additional complexities and is considered in [18, 31, 32]. The work of [18, 31, 32, 24] is focused on control of MIMO linear systems with actuator or state failures. However, for most practical applications, the linear control synthesis on FTC only guarantees stability in a region about operating point and it incurs possibly degradation in performance and instability over a large domain of operation.

The goal of this chapter is to design and analyze a fault monitoring and fault-tolerant control scheme in a real machine system. First, we construct a fault monitoring algorithm to check if the failure occurs in the system. Subsequently, the fault-tolerant control algorithm is investigated, where the neural network is used to compensate the effects of the nonlinear failures in the system. Finally, the case study is conducted based on a real-time machine actuation system. The results show that the proposed control scheme can compensate and accommodate the effects of the fault occurrences.

2. Problem Statements

Consider the following MIMO mechanical system described by

$$\left.\begin{array}{rl} x_i^{(n_i)} &= f_i(\mathbf{x},t) + \sum_{j=1}^m g_{ij}(\mathbf{x})u_j \\ &\quad + \eta_i(\mathbf{x},t) + \beta_i(t-T)\zeta_i(\mathbf{x}) \\ y_i &= x_i \end{array}\right\}, \qquad (1)$$

where $x_i^{(n_i)} = d^{n_i}x_i/dt^{n_i}$, $\mathbf{x} = [x_1, ...x_1^{(n_1-1)}, x_2, ..., x_2^{(n_2-1)}, ..., x_m, ..., x_m^{(n_m-1)}]^T \in R^n$ with $n = n_1 + n_2 + ... + n_m$, is the overall state vector, $u_i \in R, i = 1, 2, ..., m$, are the inputs and $y_i \in R, i = 1, 2, ..., m$, are the outputs of the system. The nonlinear functions $f_i, g_{ij}, i, j = 1, 2, ..., m$, are assumed to be known and the functions $\eta_i, i = 1, 2, ..., m$, represent the system uncertainties. The terms $\zeta_i, i = 1, 2, ...m$, are unknown functions which represent the faults in the system respectively, $\beta_i(t-T), i = 1, 2, ..., m$, represent the time profiles of the faults, and T is the fault-occurrence time. The system (1) can also be written in the compact form

$$\begin{aligned} x^{(n)} &= F(\mathbf{x},t) + G(\mathbf{x})u + \eta(\mathbf{x},t) + \mathcal{B}(t-T)\zeta(\mathbf{x}), \\ y &= C\mathbf{x} \end{aligned} \quad (2)$$

where
$x^{(n)} = [x_1^{(n_1)}, x_2^{(n_2)}, ..., x_m^{(n_m)}]^T \in R^m$,
$u = [u_1, u_2, ..., u_m]^T \in R^m$,
$y = [y_1, y_2, ..., y_m]^T \in R^m$, and

$$\begin{aligned} F(\mathbf{x},t) &= [f_1(\mathbf{x},t), f_2(\mathbf{x},t), ..., f_m(\mathbf{x},t)]^T, \\ G(\mathbf{x}) &= \begin{bmatrix} g_{11}(\mathbf{x}) & \cdots & g_{1m}(\mathbf{x}) \\ \vdots & \cdots & \vdots \\ g_{m1}(\mathbf{x}) & \cdots & g_{mm}(\mathbf{x}) \end{bmatrix}, \\ \eta(\mathbf{x},t) &= [\eta_1(\mathbf{x},t), \eta_2(\mathbf{x},t), ..., \eta_m(\mathbf{x},t)]^T, \\ \zeta(\mathbf{x}) &= [\zeta_1(\mathbf{x}), \zeta_2(\mathbf{x}), ..., \zeta_m(\mathbf{x})]^T, \\ C &= diag\{C_1, C_2, ..., C_m\}, \; C_i = [c_i, 0, ..., 0]_{1 \times n_i}. \end{aligned}$$

where $c_i, i = 1, ..., n_i$ are constants.

In this chapter, we consider faults with time profiles modeled by

$$\begin{aligned} \mathcal{B}(t-T) &= diag\{\beta_1(t-T), \beta_2(t-T), ..., \beta_m(t-T)\}, \\ \beta_i(t-T) &= \begin{cases} 0 & t < T \\ 1 - e^{-\theta_i(t-T)} & t \geq T \end{cases}, \end{aligned} \quad (3)$$

where the fault-occurrence time T is unknown, and $\theta_i > 0$ is an unknown constant that represents the rate at which the fault in states and actuators evolves.

The model (1) includes a large class of nonlinear robotic systems (Figure 1a and Figure 4) and injection moulding machine (Figure 1b). For example [27], most robotic systems can be described by

$$\ddot{q} = M^{-1}[\tau - V_m(q,\dot{q}) - F(\dot{q}) - G(q) - \tau_d] + \mathcal{B}(t-T)\zeta \quad (4)$$

where $q = [q_1, q_2, ...q_n]^T \in R^n$ are the joint position of the subsystem $i \in [1,n]$; $M(q)$ are the symmetric positive definite inertia matrix; $V_m(q,\dot{q})$ represent Coriolis and centripetal forces; $F(\dot{q})$ are the dynamic frictional force matrix; τ_d are a load disturbance matrix; $G(q)$ are the potential energy terms; τ denote generalized input control of the system applied at the joints. In the literature [28], the injection moulding machine has the following form

$$x^{(3)} = f(x, \dot{x}, \ddot{x}) + g(x)u \quad (5)$$

The detailed parameters can be found in [28].

The present chapter has the following two objectives: 1) It can detect a fault when the monitored system fails to function normally; 2) After a fault is detected, it is required that

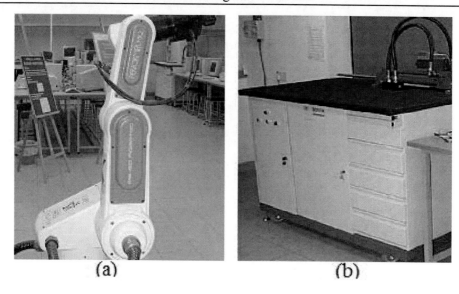

Figure 1. Robotic system and injection moulding machine.

the controller should be reconfigured to accommodate the fault. The basic assumptions for the problems stated are

A1) The fault function $\zeta(\mathbf{x})$ is uniformly continuous.
A2) The matrix $G(\mathbf{x})$ is invertible.
A3) The modeling uncertainty $\eta_i(\mathbf{x},t)$ is bounded by a known function, i.e.,

$$|\eta_i(\mathbf{x},t)| \leq \bar{\eta}_i(\mathbf{x},t), \qquad (6)$$

where the bounding function $\bar{\eta}(\mathbf{x},t)$ is continuous and uniformly bounded.

A4) The desired trajectories $y_d = [y_{d1}, y_{d2}, ..., y_{dm}]^T$ are known bounded functions of time with bounded known derivatives.

In order to be able to design the output feedback control, let us re-write system (2) as

$$\left.\begin{array}{rl}\dot{\mathbf{x}} &= A_0\mathbf{x} + b[F(\mathbf{x},t) + G(\mathbf{x})u + \eta(\mathbf{x},t) \\ & \quad + \mathcal{B}(t-T)\zeta(\mathbf{x})] \\ y &= C\mathbf{x}\end{array}\right\} \qquad (7)$$

where

$$A_0 = diag\{A_{01}, A_{02}, ..., A_{0m}\} \qquad (8)$$
$$b = diag\{b_1, b_2, ..., b_m\} \qquad (9)$$
$$A_{0i} = \begin{bmatrix} 0 & 1 & 0 & \cdots & 0 \\ 0 & 0 & 1 & \cdots & 0 \\ \vdots & \vdots & \vdots & \cdots & \vdots \\ 0 & 0 & 0 & \cdots & 0 \end{bmatrix}_{n_i \times n_i}, b_i = \begin{bmatrix} 0 \\ 0 \\ \vdots \\ 1 \end{bmatrix}_{n_i \times 1}, \qquad (10)$$
$$x = [x_1, x_2, ..., x_m]^T. \qquad (11)$$

The proposed fault control scheme makes use of the assumptions:

A5) (C, A_0) is observable.

A6) $F(\mathbf{x}, t)$ is Lipschitz in x i.e., $||F(\mathbf{x}, t) - F(\hat{\mathbf{x}}, t)|| \leq L_F ||\mathbf{x} - \hat{\mathbf{x}}||$, and $G(\mathbf{x})$ is Lipschitz in y i.e., $||G(\mathbf{x}) - G(\hat{\mathbf{x}})|| \leq L_G ||y - \hat{y}||$.

A7) $||[G(\mathbf{x}_1) - G(\mathbf{x}_2)] G^{-1}(\mathbf{x}_2)|| \leq I_G$.

A8) The modeling uncertainty is bounded by a known constant, i.e., $||\eta(\mathbf{x}, t)|| \leq \bar{\eta}$.

3. Background on Neural Networks

The focus of this section is on artificial neural networks for control purposes. Possessing certain similar characteristics as their biological counterparts, artificial neural networks have been used to learn how to control systems by observing the way that a human performs a control task, and to learn how best to control a system by taking control actions, checking the actual responses generated when these actions are used with the desired responses, then adjusting the parameters of the system so that the response of the system can be improved.

A typical two layer NN consists of two layers of tunable weights. The hidden layer has L neurons and the output layer has m neurons. The multilayer NN is a nonlinear mapping from input space into output space. Given $x \in R^n$, a two-layer NN has a net output given by

$$y = W^T \Phi(x) \tag{12}$$

where vectors $x = [x_1, x_2, ..., x_n]^T$, $y = [y_1, y_2, ..., y_m]^T$ and Φ is the nonlinear activation function.

It has been proven mathematically that any continuous functions can be uniformly approximated by a neural network. In the following sections, the neural network will be used to represent the unknown nonlinear function and achieve the control objective.

4. Fault Monitoring Algorithm

System monitoring, fault diagnosis and predictive maintenance are fast becoming key and integral components of modern production systems. An efficient diagnostic system can maintain tools in good condition and prevent severe failures by detecting and localizing faulty components at an early stage. When you come to think about it, there are only two ways you can build a fault monitoring system: add hardware sensors, design software algorithms. Traditionally, a monitoring system is achieved through the use of hardware measurement. The major problems encountered with hardware system are the extra cost and space requirement. To overcome the problems, software approaches based on signal processing or mathematical model techniques have been developed. In this section, the model-based fault monitoring algorithm and its constituent components will be elaborated in detail. The construction of a nonlinear estimated model is first designed based on the system model. Utilizing this estimation model, a time-varying threshold bound is developed so that it can serve to give a warning signal when a fault occurs.

We consider the following nonlinear model as an observer.

$$\dot{\hat{\mathbf{x}}} = A_0 \hat{\mathbf{x}} + F(\mathbf{x}, t) + G(\mathbf{x})u + L_o(y - \hat{y}), \tag{13}$$

$$\hat{y} = C\hat{\mathbf{x}} \tag{14}$$

where $\hat{\mathbf{x}}$ denotes the estimated state vector $\mathbf{x}$, L_o is a constant matrix, and $C = \begin{bmatrix} c_1 \\ c_2 \\ \vdots \\ c_m \end{bmatrix}$.

The next step in the construction of the fault detection scheme is the design of the algorithm for monitoring a fault occurrence. Based on the estimated model (13), a fault estimation algorithm is presented. Since $\mathcal{B}(t - T)\zeta(\mathbf{x})$ is zero when $t < T$, each component $\tilde{y}_i(t) = y_i - \hat{y}_i$, $i = 1, 2, ..., m$, of the state estimation error is given by

$$\tilde{y}_i(t) = c_i e^{-\bar{A}t}\tilde{\mathbf{x}}(0) + c_i \int_0^t e^{-\bar{A}(t-\tau)} b\eta(\mathbf{x}, \tau)d\tau, \quad t < T. \tag{15}$$

The time-varying threshold bound ϖ_i is chosen as follows,

$$\varpi_i = ||c_i e^{-\bar{A}t}\tilde{\mathbf{x}}(0)|| + \int_0^t ||c_i e^{-\bar{A}(t-\tau)} b|| \bar{\eta}(\mathbf{x}, \tau)d\tau, \quad t < T. \tag{16}$$

The decision that a fault has occurred is made when at least one component of the estimation error $|\tilde{y}_i(t)|$ exceeds its corresponding threshold bound ϖ_i. Figure 2 shows the block diagram of the fault monitoring.

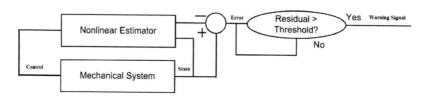

Figure 2. Block diagram of fault monitoring.

5. Fault-Tolerant Control Algorithm

Fault-tolerant control is controller designed around the concepts of fault tolerance. It is required to be able to handle several possible failures, including hardware-related faults such as system components failures, actuator device failures, or other temporary or permanent failures. Fault-tolerant scheme is typically achieved through reconfiguration of the feedback control system designed. In this section, we develop a nonlinear fault-tolerant controller and analyze the properties of the proposed scheme.

Before the fault occurrence, we first consider a controller of the system (1) when there is no fault.

For a given desired trajectory $y_{di}(t) \in R$, we define the errors $e_i(t) = y_i - y_{di}$. Then, the filtered tracking error is given by

$$\dot{s}_1 = \left(\frac{d}{dt} + k_1\right)^{n_1-1} e_1, \ldots, \dot{s}_m = \left(\frac{d}{dt} + k_m\right)^{n_m-1} e_m, \tag{17}$$

where $k_1, ..., k_m$ are positive constants to be selected. From the result of [27], it is well-known that for $s_i(t) = 0$, we have a set of linear differential equations whose solutions

$e_i, i = 1, 2, ..., m$, converge to zero with constants $(n_i - 1)/k_i, i = 1, 2, ..., m$. Thus, the system equation can be written as

$$\dot{S}(t) = F + Gu + v - y_d^{(n)} + \eta + \mathcal{B}(t - T)\zeta(\mathbf{x}), \tag{18}$$

where $S = [s_1, s_2, ..., s_m]^T$, $y_d^{(n)} = [y_{d1}^{n_1}, y_{d2}^{(n_2)}, ...y_{dn}^{(n_m)}]^T$, and $v = [v_1, v_2, ..., v_m]^T$ with $v_i = k_i^{n_i-1}\dot{e}_i + (n_i - 1)k_i^{n_i-2}\ddot{e}_i + ... + (n_i - 1)k_i e_i^{(n_i-1)}$.

In the absence of faults, the original system (18) becomes

$$\dot{S}(t) = F(\mathbf{x}, t) + G(\mathbf{x})u + v - y_d^{(n)} + \eta(\mathbf{x}, t). \tag{19}$$

and the original control law is given by

$$u = G^{-1}(\mathbf{x})[-F(\mathbf{x}, t) - v - \Lambda S], \tag{20}$$

where Λ is the same as in (13) and the parameter $\delta > 0$ is a design constant.

When a failure is detected, we need to reconfigure the feedback controller. In this case, we assume that the fault function $\zeta(\mathbf{x})$ can be approximated by a general one layer neural network (NN) (see the literature [26]) as

$$\zeta(\mathbf{x}) = W^{*T}\Phi(\mathbf{x}) + \xi, \tag{21}$$

where the bounded function approximation error ξ satisfies $||\xi|| \leq \xi_M$ with constant ξ_M and the ideal weight W^* is defined as:

$$W^* := argmin_{W \in \Omega_W}\{\sup_{\mathbf{x} \in \Omega_q} ||W^T\Phi(\mathbf{x}) - \zeta(\mathbf{x})||\}. \tag{22}$$

In general, the weights W^* are unknown and need to be estimated for controller design. Let $\hat{W}$ be estimates of the ideal W^*. Then, an estimate $\hat{\zeta}(\mathbf{x})$ of $\zeta(\mathbf{x})$ can be given by

$$\hat{\zeta}(\mathbf{x}) = \hat{W}^T\Phi(\mathbf{x}). \tag{23}$$

Therefore, the fault-tolerant control law is reconfigured by

$$u = G^{-1}(\mathbf{x})[-F(\mathbf{x}, t) - v - \Lambda S - \hat{W}^T\Phi(\mathbf{x})], \tag{24}$$

with the learning rule

$$\dot{\hat{W}} = \Upsilon\Phi(\mathbf{x})S^T - \rho\Upsilon(\hat{W} - W_a), \tag{25}$$

where $\Upsilon = \Upsilon^T > 0, \rho > 0$, and W_a is a design constant vector. Figure 3 shows the block diagram of the fault-tolerant control.

6. Output Feedback Control Design

In the preceding section, all the results have been obtained under the assumption that the full state of the system can be measured. This is not true for some cases. For example, velocity measurements are often contaminated with noise or unavailable from linear permanent magnetic motor. This will affect the performance of the fault monitoring and control. In this section, a method is introduced to remove this limitation and consider a more realistic problem where only a part of the states can be measured.

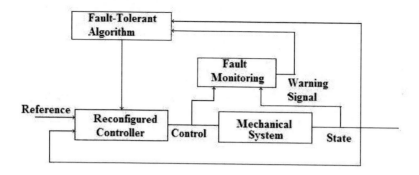

Figure 3. Block diagram of fault-tolerant control.

6.1. Fault Monitoring Algorithm

In this subsection, the fault monitoring is developed based on the output feedback control.

With the system model, the following observer is used to monitor the system state:

$$\dot{\hat{x}} = A_0\hat{x} + L_0(y - \hat{y}) + b[F(\hat{x}, t) + G(\hat{x})u], \qquad (26)$$
$$\hat{y} = C\hat{x}. \qquad (27)$$

Define the state and output estimation errors by $\tilde{x} = x - \hat{x}$ and $\tilde{y} = y - \hat{y}$ respectively. It can be easily derived that the dynamics of residual generator is governed by

$$\dot{\tilde{x}} = \bar{A}\tilde{x} + b[G(x)u - G(\hat{x})u + F(x,t) - F(\hat{x},t)]$$
$$+ b\eta(\bar{x}, t), \qquad (28)$$
$$\tilde{y} = C\tilde{x}, \qquad (29)$$

where $\bar{A} = A_0 - L_0 C$. The gain matrix L_0 is chosen so that $\bar{A}$ is stable. We consider the Lyapunov function $V_0 = \tilde{x}^T P \tilde{x}$ and its derivative is given by

$$\dot{V}_0 = \tilde{x}^T(\bar{A}^T P + P\bar{A})\tilde{x} + 2\tilde{x}^T Pb[G(x)u - G(\hat{x})u]$$
$$+ 2\tilde{x}^T Pb[F(x,t) - F(\hat{x},t)] + 2\tilde{x}^T Pb\eta(x,t). \qquad (30)$$

Note that the last three terms satisfy the inequalities

$$2\tilde{x}^T Pb[G(x)u - G(\hat{x})u] \le \tilde{x}^T P^2 \tilde{x} + I_G^2 ||b||^2 ||u_0||^2,$$
$$2\tilde{x}^T Pb[F(x,t) - F(\hat{x},t)] \le \gamma \tilde{x}^T P^2 \tilde{x} + \gamma^{-1} ||b||^2 L_F^2 ||\tilde{x}||^2,$$
$$2\tilde{x}^T Pb\eta(x,t) \le ||\tilde{x}||^2 + ||Pb||^2 \bar{\eta}^2,$$

where we have used Assumption **A 7**, $u = G^{-1}(\hat{x})u_0$ and $\gamma = ||b||^2 L_F^2$. This implies that

$$\dot{V}_0 \le \tilde{x}^T(\bar{A}^T P + P\bar{A} + (||b||^2 L_F^2 + 1)P^2 + I)\tilde{x}$$
$$+ I_G^2 ||b||^2 ||u_0||^2 + ||Pb||^2 \bar{\eta}^2$$
$$\le -\tilde{x}^T Q\tilde{x} + I_G^2 ||b||^2 ||u_0||^2 + ||Pb||^2 \bar{\eta}^2$$
$$\le -\frac{\lambda_{min}(Q)}{\lambda_{max}(P)} V_0 + I_G^2 ||b||^2 ||u_0||^2 + ||Pb||^2 \bar{\eta}^2. \qquad (31)$$

Fault Monitoring and Control of Mechanical Systems

In the above analysis, it is assumed that

$$\bar{A}^T P + P\bar{A} + (||b||^2 L_F^2 + 1)P^2 + I + Q \leq 0 \tag{32}$$

By using *Lemma 3.2.4* of [33], we have

$$\mathcal{V}_0 \leq e^{-\frac{\lambda_{min}(Q)}{\lambda_{max}(P)}t} \mathcal{V}_0(0) + \pi(t). \tag{33}$$

where $\pi(t) = \int_0^t e^{-\frac{\lambda_{min}(Q)}{\lambda_{max}(P)}(t-\tau)}[I_G^2 ||b||^2 \, ||u_0||^2 + ||Pb||^2 \bar{\eta}^2] d\tau$.
Using $||\tilde{x}||^2 \leq \frac{1}{\lambda_{min}(P)} \mathcal{V}_0$, one obtains

$$||\tilde{y}|| \leq ||C|| \sqrt{\frac{e^{-\frac{\lambda_{min}(Q)}{\lambda_{max}(P)}t}}{\lambda_{min}(P)} \mathcal{V}_0(0) + \frac{\pi(t)}{\Lambda_{min}(P)}} = \varpi \tag{34}$$

In the above result, $\mathcal{V}_0(0)$ can be replaced by a conservative estimate δ_0, where $|\mathcal{V}_0(0)| \leq \delta_0$. Since the first term contains an exponential function $e^{-\frac{\lambda_{min}(Q)}{\lambda_{max}(P)}t}$, the replacement will not affect the threshold seriously. The fault detection can be carried out as

$$\begin{cases} ||\tilde{y}|| \leq \varpi, & \text{no fault occurs} \\ ||\tilde{y}|| > \varpi, & \text{fault has occurred} \end{cases} \tag{35}$$

6.2. Fault-Tolerant Control

When detecting a fault or receiving a warning signal from the monitoring system, the fault tolerance in the control system can be achieved through adding a NN approximator into the normal controller. In this section, we use the same control policy as shown in state feedback form to compensate the effects of the faults.

Define the state error $E = \mathbf{x} - \mathbf{x}_d$ where $\mathbf{x}_d = [y_{d1}, ... y_{d1}^{(n_1-1)}, y_{d2}, ..., y_{d2}^{(n_2-1)}, ..., y_{dm}, ..., y_{dm}^{(n_m-1)}]^T \in R^n$. System (7) may be expressed as

$$\begin{aligned} \dot{E} &= AE + b[F(\mathbf{x},t) + G(\mathbf{x})u - y_d^{(n)} + KE + \eta(\mathbf{x},t) \\ &\quad + \mathcal{B}(t-T)\zeta(\mathbf{x})] \end{aligned} \tag{36}$$

$$e = CE, \tag{37}$$

where $A = A_0 - bK$, $y_d^{(n)} = [y_{d1}^{(n_1)}, ..., y_{dm}^{(n_m)}]^T$ and $e = y - y_d$. The constant matrix K in (36) is chosen so that $C[sI - A]^{-1}b$ is strictly positive real (SPR).

The developed fault-toleranct control is given as follows:

$$\begin{aligned} u &= G^{-1}(\hat{\mathbf{x}})[u_0 + u_c], \\ u_0 &= -K_e e - F(\hat{\mathbf{x}},t) + y_d^{(n)}, \\ u_c &= -\hat{W}^T \Phi(\hat{\mathbf{x}}) \end{aligned} \tag{38}$$

with adaptive law $\dot{\hat{W}} = \Upsilon \Phi(\hat{\mathbf{x}}) \tilde{y}^T - \rho \Upsilon(\hat{W} - W_a)$. The reconfigured observer is given by

$$\begin{aligned} \dot{\hat{\mathbf{x}}} &= A_0 \hat{\mathbf{x}} + L(y-\hat{y}) + b \Big\{ F(\hat{\mathbf{x}},t) + G(\hat{\mathbf{x}})u + \Xi_1 + \frac{1}{2}\tilde{y} \\ &\quad + sgn(\tilde{y}) I_G ||u_c||_1 + \hat{W}^T \Phi(\hat{\mathbf{x}}) + \frac{2L_G^2 ||\hat{W}^T \Phi(\hat{\mathbf{x}})||^2 ||\tilde{y}||^2}{\lambda_e} \Big\} \end{aligned}$$

The resulting observation error equation is

$$\dot{\tilde{\mathbf{x}}} = \bar{A}\tilde{\mathbf{x}} + b\{G(\mathbf{x})u - G(\hat{\mathbf{x}})u + F(\mathbf{x},t) - F(\hat{\mathbf{x}},t) \\ + \eta - \Xi_1 - sgn(\tilde{y})I_G\|u_c\|_1 + \mathcal{B}(t-T)\zeta \\ -\hat{W}^T\Phi(\hat{\mathbf{x}}) - \frac{1}{2}\tilde{y} - \frac{2L_G^2}{\lambda_e}\|\hat{W}^T\Phi(\hat{\mathbf{x}})\|^2\|\tilde{y}\|^2\} \quad (39)$$

Utilizing the Lyapunov function and SPR condition, it is shown that the state error $\tilde{\mathbf{x}}$ and E of the closed-loop system are uniformly ultimately bounded (UUB) under the fault-tolerant control (38).

6.3. Further Extensions

In this subsection we will discuss the extension of our previous results to the case where faults and uncertainties intervene in the output equation. For this purpose, it is assumed that $y = C\mathbf{x} + \eta_y(\mathbf{x},t) + \mathcal{B}(t-T)\zeta_y(\mathbf{x})$ with $\|\eta_y(\mathbf{x},t)\| \leq \bar{\eta}_y$. The fault monitoring (26),(27) can still be used and the error equations become as

$$\dot{\tilde{\mathbf{x}}} = \bar{A}\tilde{\mathbf{x}} + b[G(\mathbf{x})u - G(\hat{\mathbf{x}})u] + b[F(\mathbf{x},t) - F(\hat{\mathbf{x}},t)] + b\eta(\mathbf{x},t) + L_0\eta_y(\mathbf{x},t) \quad (40)$$
$$\tilde{y} = C\tilde{\mathbf{x}} + \eta_y(\mathbf{x},t) \quad (41)$$

Following a similar procedure as in Section 6.1, the threshold of the monitoring in this case is given by

$$\varpi = \|C\|\sqrt{\frac{e^{-\frac{\lambda_{min}(Q)}{\lambda_{max}(P)}t}}{\lambda_{min}(P)}\mathcal{V}_0(0) + \int_0^t \frac{e^{-\frac{\lambda_{min}(Q)}{\lambda_{max}(P)}(t-\tau)}}{\lambda_{min}(P)}G_h d\tau} + \bar{\eta}_y$$

where $G_h = L_G^2\|b\|^2 \|\tilde{y}\|^2 \| u \|^2 + 2\|Pb\|^2\bar{\eta}^2 + 2\|PL\|^2\bar{\eta}_y^2$. For the fault-tolerant control, the control form such as (38) can be used.

7. Case Study

To illustrate the effectiveness of the proposed method, real-time experiments are carried out on a precision 3D Cartesian robotic system, as shown in Figure 4. Every axis of the robot is driven by a linear electric motor manufactured by Anorad Co., USA. The dSPACE control development and rapid prototyping system, in particular, the DS1103 board, is used. dSPACE integrates the whole development cycle seamlessly into a single environment. MATLAB/Simulink can be directly used in the development of the dSPACE real-time control system.

Let

$$a = K_f K_e/R, \quad (42)$$
$$b = \frac{K_f}{R}, \quad (43)$$
$$f_{fric} = \frac{R}{K_f}\bar{f}_{fric}. \quad (44)$$

Thus, we have the following equivalent model:

$$\ddot{x} = -a\dot{x} + bu - f_{fric}. \tag{45}$$

The dominant model of $\ddot{x} = -a\dot{x} + bu$, it is obtained from step test. Figures 5-6 show the step responses of actual and identified models, respectively. The model error is due to the nonlinear terms in the system. We consider the Coulomb friction by using a trial and error method. It is observed that $u = 2$ is critical point for the friction. Thus, the following model is used

$$\ddot{x} = -1.956\dot{x} + 94.714(u - 2\text{sgn}(\dot{x})) - f_{non} \tag{46}$$

The nonlinear function f_{non} is the remaining nonlinear uncertainty and bounded by $40[|\dot{x}| + 30e^{-(\dot{x}/70)^2}]$. Let $x_1 = x$ and $x_2 = \dot{x}$. The residual generator is given by

$$\dot{\hat{x}}_1 = \hat{x}_2 + k_1(y - \hat{y})$$
$$\dot{\hat{x}}_2 = -1.956\hat{x}_2 + 94.714(u - 2\text{sgn}(x_2)) + k_2(y - \hat{y})$$

where $\hat{x}_1$ and $\hat{x}_2$ are the estimates of x_1 and x_2 respectively, y and $\hat{y}$ are the actual and observer outputs respectively. The parameters k_1 and k_2 are chosen as 200 and 80 respectively such that the system is stable. The error signal $\bar{e} = y - \hat{y}$ is used for the residual generator. From the formula (16), the threshold value of the residual signal can be derived as

$$|\bar{e}| \leq \frac{1}{197.2344} \int_0^t \left[e^{-2.3608(t-\tau)} - e^{-199.5952(t-\tau)} \right] \times 40 \left[|\dot{x}| + 30e^{-(\dot{x}/70)^2} \right] d\tau \tag{47}$$

In the first experiment, the linear motor follows a trajectory $y_d = 20\sin(2t)$mm and we consider a mechanical fault due to obstruction from the cable protection chain which consists of many chain links (see Figure 7). Due to prolonged high speed operation, one or several of the chain links are jammed which obstruct the motor movement. Figure 8 shows the response due to the fault occurrence. The threshold bound can be computed from (47). Since the desired speed $\dot{y}_d = 40\cos(2t)$mm/s and the tracking control ensure that the actual speed follows the desired trajectory, the threshold value can be given by

$$|\bar{e}| \leq 0.0021 \times 40 \times [40 + 30e^{-(40/70)^2}], \tag{48}$$

i.e.,5.1780. It is observed that the residual signal exceeds the threshold value and the control performance is degraded. In this case, the fault accommodation control is triggered to compensate the effects of the fault. Figure 9 shows the variation of the tracking error, while Figures 10-11 show the compensator signal and actual control signal. It is observed that the control performance is improved following the adaptive learning.

In the second experiment, the linear motor follows a trajectory $y_d = 20\sin(6t)$mm and we consider a fault which is change in loading. An increase in static friction due to prolonged usage or inadequate maintenance has a similar effect to an increase in load. See Figure 12. an addition load is added when the linear motor is working. The residual threshold bound is given by

$$|\bar{e}| \leq 0.0021 \times 40 \times [120 + 30e^{-(120/70)^2}], \tag{49}$$

ie., 10.2134. Keeping all design parameters as in the first experiment, we apply the fault detection scheme (13) to this case. The fault detection response is shown in Figure 13. The threshold value is exceeded and the fault accommodation is triggered to compensate the effects of the fault occurrence after a learning period. Figure 14 shows the tracking error, while Figures 15-16 show the compensator signal and actual control signal.

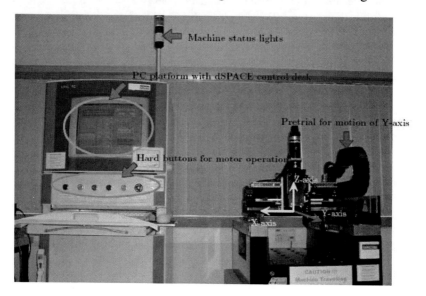

Figure 4. Control testbed.

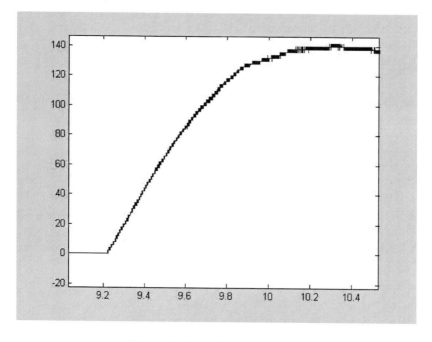

Figure 5. Actual step response.

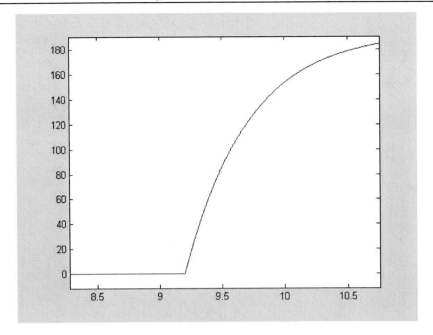

Figure 6. Model response.

Figure 7. Cable protection chain.

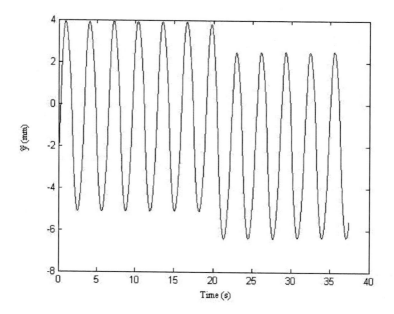

Figure 8. Residual signal when fault occurrence for case 1.

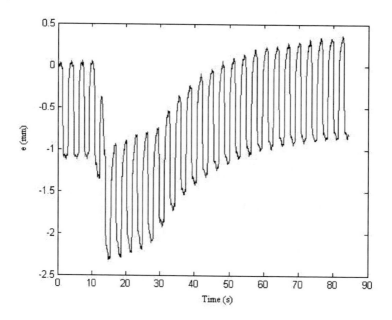

Figure 9. Tracking error with control compensation for case 1.

Fault Monitoring and Control of Mechanical Systems 225

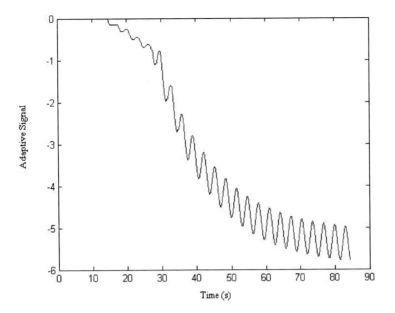

Figure 10. Adaptive signal with control compensation for case 1.

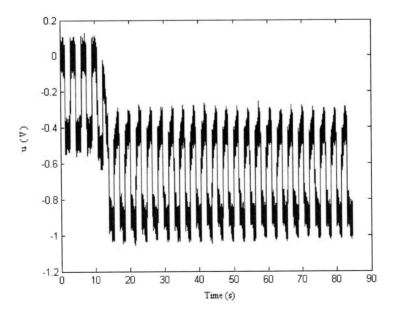

Figure 11. Control signal with control compensation for case 1.

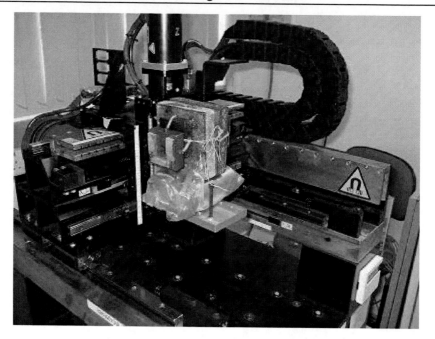

Figure 12. Loading changes.

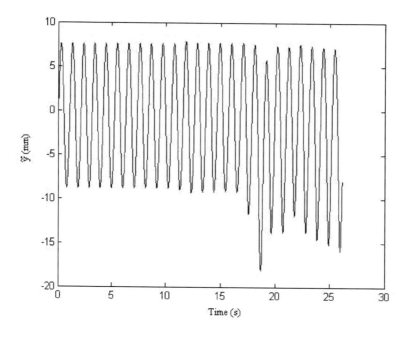

Figure 13. Residual signal when fault occurrence for case 2.

Fault Monitoring and Control of Mechanical Systems 227

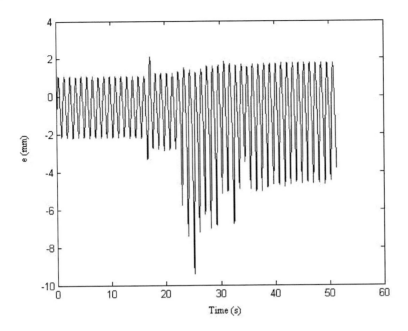

Figure 14. Tracking error with control compensation for case 2.

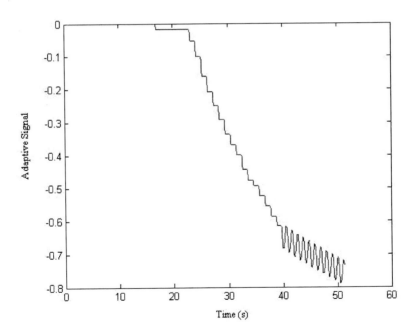

Figure 15. Adaptive signal with control compensation for case 2.

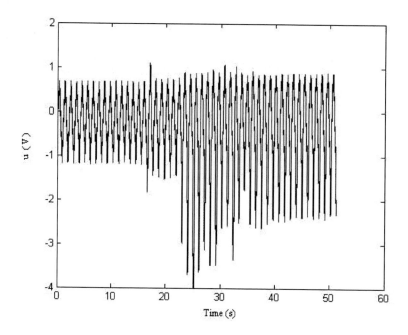

Figure 16. Control signal with control compensation for case 2.

8. Conclusions

In this chapter, the fault detection and accommodation schemes have been proposed for MIMO mechanical systems. Using an online approximation approach, we have been able to relax the parametric fault requirements of traditional adaptive control without considering the dynamic uncertainty as part of the fault. Experimental results have shown that the designed control scheme with neural network can detect and accommodate faults effectively in a motor stage. The advantages of the proposed algorithm with NN is that the neural network learning laws can guarantee the stability of the robotic systems and it is not necessary to give an off-line training for the NN control.

References

[1] Isermann, R. "Process fault detection based on modeling and estimation methods-a survey," *Automatica*, vol.20, pp387-404, 1984

[2] M.L.Visinsky, J.R.Cavallaro, and I.D.Walker, "Expert system framework for fault detection and fault olerance in robotics," Computers and Electr.Eng., vol.20,pp421-435,1994

[3] X.Q.Liu, H.Y.Zhang,J.Liu and J.Yang, " Fault detection and diagnosis of permenent-magnet DCmotor based on parameter estimation and neural network," *IEEE Trans.on Industrial Electronics,* vol.47, pp1021-1030, 2000

[4] O. Bhattacharyya, M. B. Jun, S. G. Kapoor and R. E. DeVor, "The effects of process faults and misalignments on the cutting force system and hole quality in reaming," *International Journal of Machine Tools & Manufacture*, vol.46, pp1281-1290, 2006

[5] T. Boutros and M.Liang, "Mechanical fault detection using fuzzy index fusion," *International Journal of Machine Tools & Manufacture*, vol.47, pp1702-1714, 2007

[6] H.Wang and S.Daley, "Actuator fault diagnosis: An adaptive observer-based technique," *IEEE Trans.Automat.Contr.*, vol.41, pp.1073-1078, 1996

[7] J.Wunnenberg and P.M.Frank, "Dynamic model based incipient fault detection concept for robotics," in Proc. IFAC 11th Triennial World Congr., Tallinn, Estonia, pp61-66, 1990

[8] S.H.Kia, H. Henao, and G.-A. Capolino, "A High-Resolution Frequency Estimation Method for Three-Phase Induction Machine Fault Detection," *IEEE Trans.on Industrial Electronics*, vol.54, pp2305 - 2314, 2007

[9] S. N. Huang, K. K. Tan, Y. S. Wong, C. W. de Silva, H. L. Goh and W. W. Tan, "Tool wear detection and fault diagnosis based on cutting force monitoring," *International Journal of Machine Tools & Manufacture*, vol.47, pp444-451, 2007

[10] A.T.Vemuri and M.M.Polycarpou, "Neural-network-based robust fault diagnosis in robotic systems," *IEEE Trans.on Neural Networks*, vol.8, 1410-1419, 1997

[11] B.Li, M.Y.Chow, Y.Tipsuwan, and J.C.Hung, "Neural-network-based motor rolling bearing fault diagnosis," *IEEE Trans.on Industrial Electronics*, vol.47, pp1060-1069, 2000,

[12] B.M.Wilamowski and O. Kaynak, " Oil well diagnosis by sensing terminal characteristics of the induction motor," *IEEE Trans.on Industrial Electronics*, vol.47, pp1100 - 1107, 2000

[13] T.Renato, H.T. Marco and B.Marcel, "Fault detection and isolation in cooperative manipulators via aritificial neural networks," Proc.of the IEEE International Conference on Control Applications, Mexico, pp492-497, 2001

[14] S.Wu and T.W.S.Chow, "Induction machine fault detection using SOM-based RBF neural networks," *IEEE Trans.on Industrial Electronics*, vol.51, pp183-194, 2004

[15] W.W.Tan and H.Huo, "A generic neurofuzzy model-based approach for detecting faults in induction motors," *IEEE Trans.on Industrial Electronics*, vol.52, pp1420 - 1427, 2005

[16] S.Rajagopalan, J.M. Aller, J.A. Restrepo, T.G. Habetler, and R.G. Harley, "Analytic-Wavelet-Ridge-Based Detection of Dynamic Eccentricity in Brushless Direct Current (BLDC) Motors Functioning Under Dynamic Operating Conditions," *IEEE Trans.on Industrial Electronics*, vol.54, pp1410-1419, June 2007

[17] M.S.Ballal, Z.J. Khan, H.M.Suryawanshi, and R.L. Sonolikar, "Adaptive Neural Fuzzy Inference System for the Detection of Inter-Turn Insulation and Bearing Wear Faults in Induction Motor," *IEEE Trans.on Industrial Electronics,* vol.54, pp250-258, 2007

[18] J.A.Farrell, T.Berger, and B.D.Appleby, "Using learning techniques to accommodate unanticipated faults," IEEE Contr.Syst. Mag., vol.13,pp40-49, 1993

[19] G.Tao, X.L.Ma, and S.M.Joshi, "Adaptive state feedback control of systems with actuator failures," University of Virginia, Charlottesville, VA, Tech. Rp.UVA-EE-ASC-M-990 701,1999.

[20] Y.Diao and K.M.Passino, "Stable fault-tolerant adaptive fuzzy/neural control for a turbine engine," *IEEE Trans.on Contr.Syst.Technol.*, vol.9, pp.494-509, 2001

[21] S.Chen, G.Tao and S.M.Joshi, "On matching conditions for adaptive state tracking control of systems with actuator failures," *IEEE Trans. on Automat.Contr.,* vol.47, pp.473-478,2002

[22] M.L.Visinsky, J.R.Cavallaro, and I.D.Walker, "A dynamic fault tolerance framework for remote robots," *IEEE Trans.on Robotics and Automation*, vol.11, pp477-490, 1995

[23] M.M.Polycarpou and A.J.Helmicki, "Automated fault detection and accommodation:A learning systems approach," IEEE Trans.on Syst., Man Cybern., vol.25,pp1447-1458,1995

[24] R.L.A.Ribeiro, C.B.Jacobina,E.R.C.da Silva and A.M.N.Lima, "Fault-tolerant voltage-fed PWM inverter AC motor drive systems," *IEEE Trans.on Industrial Electronics,* vol.51, pp439-446, 2004

[25] J.Park and I.W.Sandberg, "Universal approximation using radial basis function networks," *Neural Computation*, vol.3, pp 246-256, 1991

[26] S.N.Huang,K.K.Tan and T.H.Lee,"A decentralized control of interconnected systems using neural networks," *IEEE Trans.on Neural Networks*, vol.13, no.6,pp.1554-1557, 2002.

[27] J.J.E.Slotine and W.Li, *Applied Nonlinear Control*. Upper Saddle River, NJ:Prentice-Hall,1991

[28] K.K.Tan,S.N.Huang,and X.Jiang, "Adaptive control of ram velocity for the injection moulding machine," *IEEE Trans.on Control Systems Technology,* vol.9, pp663-671, 2001

[29] C.Y.Huang and R.F.Stengel, "Restructable control using proportional-integral implicit model following," *J. Guidance, Contr., Dyn.,* vol.13, no.2, pp303-309,1990

[30] P.V.Goode, and M.Y. Chow, "Using a neural/fuzzy system to extract heuristic knowledge of incipient faults in induction motors. Part I-Methodology," *IEEE Trans.on Industrial Electronics,* vol.42, pp131 - 138,1995

[31] M.L.Visinsky, J.R.Cavallaro, and I.D.Walker, "A dynamic fault tolerance framework for remote robots," *IEEE Trans.Robotics and Automation*, vol.11, pp477-490, 1995

[32] S.Chen, G.Tao and S.M.Joshi, "On matching conditions for adaptive state tracking control of systems with actuator failures," *IEEE Trans.Automat.Contr.,* vol.47, pp.473-478, 2002

[33] P.A.Ioannou and J.Sun, *Robust Adaptive Control*, Prentice Hall. Englewood Cliffs, NJ, 1995.

REVIEWED by
Professor Cong Ming
Key Laboratory for Precision and Non-traditional Machining Technology of Ministry of Education,
School of Mechanical Engineering,
Dalian University of Technology
Dalian, 116023,
P.R.China

INDEX

A

accelerometers, 47
accurate models, viii, 39
actuation, ix, 111, 153, 155, 172, 212
actuators, vii, x, 80, 83, 109, 156, 211, 212, 213
adaptation, 24
aluminium, 104
amplitude, 43
anchoring, 40
anisotropy, 165
assessment, 11, 212
asymmetry, 91
atomic force microscope (AFM), 138, 191
Austria, 191
automate, 194
automation, 13, 191, 210

B

balanced systems, 80
bandwidth, 47, 111
barriers, 188
base, 24, 115, 118, 156, 161, 192, 195
Beijing, 153
Belgium, 79, 172
bending, 47, 49, 56, 97, 98, 99, 110
benefits, 63, 115
boreholes, 177, 178
brass, 189, 190
Brazil, 1
breakdown, 6, 13
burn, 8
business model, 115

C

calibration, 12
CAM, x, 37, 117, 193, 194, 206, 210
carbon, 19, 20
case study, 33, 35, 119, 130, 131, 212
casting, 12, 188
chemical, 16, 201
China, 115, 153, 231
classification, 212
coding, 119
collisions, 108
Colombia, 1
commercial, 154, 189, 190
compatibility, 122
compensation, 201, 224, 225, 227, 228
complexity, x, 24, 31, 187
compliance, 118
computer, viii, x, 117, 155, 193, 194, 201, 208
computer-aided design (CAD), x, 37, 117, 193, 194, 200, 201, 206, 208, 210
conditioning, 159, 166
configuration, 9, 23, 25, 40, 69, 74, 75, 157, 159
construction, 114, 156, 215, 216
consumption, viii, 81, 104, 105, 108, 109, 111, 112, 113, 114
contour, x, 129, 169, 193, 205, 206, 208
convergence, 53, 63, 64
correlation, 51, 52, 53, 78, 79, 181
correlation coefficient, 79
corrosion, 9
cost, x, 3, 24, 32, 59, 63, 65, 134, 135, 136, 137, 187, 188, 190, 215
covering, viii, 39, 46
CPT, 176
cutting force, 87, 90, 118, 119, 129, 130, 132, 139, 143, 144, 148, 190, 211, 229

D

damping, 48, 50, 56, 59, 60, 61, 65, 79, 90, 92, 93, 94, 98, 99, 100, 102, 104, 107, 108, 109, 110, 111, 113, 114, 116, 121, 128, 202
data set, 179
database, 23, 51
deaths, 32
decay, 202
decision-making process, vii, 1
decomposition, 24
defects, 8, 16, 23, 24, 30, 32
deficiencies, 2
deformation, 16, 84, 88, 201
degradation, 3, 23, 212
dematerialization, 83
dependent variable, 143
deposition, 188, 190
depth, 6, 16, 87, 89, 91, 93, 94, 95, 98, 99, 109, 110, 111, 112, 118, 119, 129, 130, 133, 134, 135, 136, 137, 139, 143, 144, 146, 177, 178
derivatives, 60, 63, 214
designers, 29, 117, 154
detection, 3, 222, 228, 229
deviation, 13, 140, 142, 203, 204
differential equations, 42, 59, 216
direct costs, 211
dispersion, 17, 194
displacement, 40, 48, 56, 62, 65, 67, 88, 89, 90, 94, 99, 121, 125
distribution, vii, 2, 7, 8, 9, 12, 13, 15, 16, 17, 18, 19, 20, 21, 22, 23, 34, 35, 37, 59, 146, 170
distribution function, 16, 22
drawing, 33

E

economic growth, 187
electromagnetic, 98
electron, 188
EMS, 187, 190
end-users, 84
energy, viii, 59, 81, 82, 83, 104, 108, 109, 111, 112, 113, 114, 188, 189, 213
energy consumption, viii, 81, 104, 108, 109, 111, 112, 114
energy efficiency, 112
engineering, 15, 37, 176
environment, 12, 13, 184, 188, 211, 220
environmental conditions, 2, 6, 13
environmental impact, viii, 81, 82, 83, 114
environmental stress, 12

environmental stresses, 12
equality, 64
equilibrium, 42, 122
equipment, 4, 6, 7, 8, 9, 11, 13, 15, 36, 190
Estonia, 229
evolution, 18, 74, 75, 77
excitability, 55
excitation, 48, 54, 55, 56, 57, 58, 63, 65, 67, 70, 85, 86, 90, 109
execution, 21, 25
experimental condition, 20
extracts, 197

F

fabrication, 188, 190
factories, 191
fault detection, 211, 212, 216, 219, 222, 228, 229, 230
fault diagnosis, 212, 215, 229
fault tolerance, 216, 230, 231
FEM, 79, 101, 102, 103
field theory, 176
financial support, 148
flank, 16, 17, 19, 20, 21, 22, 131, 139, 144
flaws, 29, 30
flexibility, 63, 70, 99, 102, 107, 108, 109, 111
force, 30, 47, 55, 57, 59, 65, 84, 89, 94, 98, 105, 118, 119, 120, 121, 122, 123, 125, 129, 130, 138, 139, 140, 142, 143, 144, 145, 148, 155, 159, 160, 161, 162, 163, 164, 165, 190, 191, 194, 195, 196, 197, 201, 202, 203, 204, 205, 206, 207, 208, 213
formation, 16
formula, 221
France, 115
freedom, viii, 39, 42, 92, 93, 94, 99
friction, 16, 40, 59, 176, 221
fusion, 211, 229

G

geometrical parameters, 166, 167, 168
geometry, 5, 16, 30, 32, 47, 48, 85, 118, 121
Germany, 185
grain size, x, 193, 194
gratings, 189
gravitational force, 165
grouping, 5, 24, 66
growth, 29, 31, 73, 75, 141, 142, 187
growth rate, 73, 75, 142
Guangzhou, 115

H

hardness, 143
health care, 187
height, 40, 100, 156, 167
host, 188
housing, 108
human, 11, 12, 13, 32, 117, 215
hybrid, x, 156, 173, 187, 188, 194

I

images, 21, 139, 140
immersion, 85, 102, 113, 135, 139
improvements, 54
incidence, 85
independent variable, 143
India, ix, 175, 176, 178
induction, 229, 230
industries, vii, 1, 36
industry, vii, ix, 2, 11, 12, 153, 154, 171, 190
inertia, 53, 104, 154, 159, 160, 161, 162, 163, 164, 165, 166, 167, 172, 202, 213
infant mortality, 23
injury, 11
inspectors, 23
integration, 114
integrity, 188
interface, 23, 117, 196
interference, 77, 167
inversion, 185, 191
investment, 188
isolation, 229
issues, 118, 119
Italy, 191
iteration, 53

J

Japan, 192, 210
joints, vii, x, 53, 121, 165, 191, 211, 213

K

Korea, 191

L

laws, 228
lead, 12, 22, 40, 46, 84, 87, 93, 96, 119, 165

learning, ix, 175, 176, 184, 185, 217, 221, 222, 228, 230
LED, x, 193, 194, 195, 205, 206, 207, 208, 209, 210
lens, x, 193, 194, 195, 205, 206, 207, 208, 209
light, 12, 104, 105, 106, 107, 108, 109, 111, 112, 115, 130, 134, 146
light conditions, 12
linear systems, 80, 212
lithography, x, 187, 190
localization, 49
locus, 73, 74
Lyapunov function, 218, 220

M

machine learning, ix, 175, 176, 184
machinery, 212
magnet, 211, 228
magnitude, vii, 2, 8, 22, 55, 61, 77
majority, 82, 84, 104
manipulation, 188
manufacturing, vii, x, 1, 2, 3, 4, 5, 8, 11, 12, 13, 15, 16, 21, 22, 23, 24, 25, 26, 27, 28, 29, 30, 31, 32, 34, 35, 36, 37, 38, 115, 117, 187, 188, 190, 198, 201, 205
mapping, 158, 180, 185, 215
mass, x, 42, 43, 44, 50, 53, 86, 88, 93, 94, 98, 104, 105, 106, 107, 108, 114, 120, 128, 160, 161, 163, 164, 167, 188, 190, 193, 194
material resources, 83, 104, 114
materials, 23, 30, 60, 85, 86, 98, 104, 188, 189, 190
matrix, 44, 49, 50, 54, 59, 60, 61, 62, 65, 71, 88, 89, 91, 122, 125, 154, 157, 158, 159, 160, 161, 165, 166, 167, 179, 213, 214, 216, 218, 219
measurements, 12, 30, 49, 51, 65, 69, 78, 84, 120, 121, 126, 127, 129, 140, 176, 201, 202, 217
mechanical performances, 83
mechanical properties, 32, 187
mechanical testing, 191
medical, 187
metals, 188
methodology, 2, 16, 33, 77, 82, 100, 114, 155, 172, 179
Mexico, 173, 189, 229
microfabrication, x, 187, 190
micrometer, 189, 190
microscope, 21, 138, 139, 142, 191
microstructures, 188
miniature, 189, 190
miniaturization, ix, x, 187, 190
model reduction, 70, 80

modelling, 41, 49, 53, 58, 59, 60, 79
models, viii, 12, 34, 39, 41, 49, 51, 56, 59, 93, 115, 118, 120, 121, 126, 127, 128, 148, 180, 184, 211, 221
modules, 172
modulus, 88, 128, 176
mold, 194
mortality, 23
moulding, 213, 214, 230
multimedia, 202

N

Netherlands, 79, 115
neural network, ix, 175, 176, 185, 212, 215, 228, 229, 230
neural networks, 215, 229, 230
neurons, 178, 179, 215
next generation, 190
nodes, 42, 51, 65
normal distribution, 35
numerical analysis, 11

O

obstruction, 221
oil, x, 15, 193, 194
operating costs, 115
operating range, 12
operations, vii, 1, 3, 22, 23, 24, 25, 26, 33, 34, 35, 36, 39, 40, 84, 109, 113, 119, 211
optimization, viii, 39, 155, 167, 173, 180, 189
orthogonality, 44, 60, 62
overtraining, 182

P

pairing, 51, 52
parallel, ix, 9, 11, 94, 98, 99, 104, 153, 154, 155, 156, 158, 159, 165, 167, 169, 171, 172, 173, 191, 195
parameter estimation, 53, 211, 228
Pareto, 14
path planning, 194
pattern recognition, 185
permit, 188
Philadelphia, 150
physical properties, 49
physics, 118
platform, 156, 157, 158, 159, 160, 163, 164, 165, 188

plausibility, 180
polar, 102, 111, 112
policy, 13, 219
polymer, 188
positive feedback, 90
power plants, 7
prevention, 15
probability, vii, viii, 2, 6, 7, 8, 9, 11, 15, 16, 17, 18, 22, 28, 31, 32, 117, 148
probability density function, 9
probability distribution, 6, 7, 16, 18, 22
product design, 2, 4
profit, 118, 134
programming, viii, 117, 118, 196
project, 176, 188
protection, 221, 223
prototype, ix, 105, 106, 114, 153, 154, 171, 189, 192

Q

quality control, 3, 4, 8, 23, 32
quality of life, 187

R

radius, 16, 154, 172, 205, 206
random errors, 13
reciprocity, 126
rectification, 211
redundancy, ix, 9, 153, 155, 172
regeneration, 40, 73, 82, 85, 118
regression, ix, 139, 140, 144, 175, 176, 180, 185
reliability, 1, 2, 3, 4, 6, 7, 8, 9, 11, 12, 13, 15, 16, 17, 18, 19, 20, 21, 22, 23, 24, 25, 26, 27, 28, 29, 32, 34, 35, 36, 37, 81, 84, 118
repair, 15, 31
reprocessing, 32
requirements, vii, x, 1, 5, 12, 31, 34, 83, 84, 86, 118, 176, 187, 228
resistance, 185
resolution, 2, 32, 73, 95, 114, 189, 194, 195, 196, 208
resources, vii, 1, 4, 83, 104, 114, 188
response, viii, 39, 41, 42, 43, 45, 46, 57, 58, 59, 63, 64, 65, 67, 70, 71, 73, 109, 117, 118, 119, 120, 121, 125, 126, 144, 148, 215, 221, 222, 223
restrictions, 4
revenue, 134
risks, 115
robotics, 154, 228, 229

roots, 73, 74, 75, 179
rotations, 122
roughness, 29, 30, 31, 33, 139, 147
rubber, 201
rules, 24, 82

S

safety, vii, viii, x, 11, 117, 118, 120, 130, 133, 134, 137, 138, 144, 148, 187, 211
scaling, 155, 188
scope, 179
sensing, 229
sensitivity, 53, 190
sensors, vii, x, 13, 98, 211, 215
sequencing, 24, 25
shape, x, 7, 8, 43, 45, 52, 56, 61, 193, 194, 200, 201, 205, 206, 208
shear, 176
showing, 77, 78, 102, 177
signals, 45, 119
silicon, 188
simulation, 41, 39, 59, 69, 75, 76, 77, 125, 126, 142, 144, 155
Singapore, 211
smoothing, 185
social environment, 12
software, 117, 154, 195, 207, 215
solution, 43, 59, 63, 64, 65, 86, 90, 91, 154
solution space, 154
Spain, 39, 81, 115
specifications, 4, 23, 34, 84, 211
spindle, vii, viii, 2, 23, 30, 53, 85, 91, 101, 117, 118, 119, 120, 121, 124, 125, 126, 127, 129, 130, 131, 132, 133, 134, 135, 137, 139, 142, 143, 144, 145, 146, 148, 188, 201
stability, viii, 29, 73, 75, 78, 81, 86, 91, 93, 94, 95, 98, 102, 108, 109, 117, 118, 119, 129, 133, 137, 138, 144, 146, 148, 188, 212, 228
standard deviation, 13, 140, 142
state, viii, 39, 59, 60, 66, 67, 68, 69, 70, 71, 72, 78, 118, 212, 216, 217, 218, 219, 220, 230, 231
states, 66, 70, 71, 213, 217
steel, 19, 20, 21, 33, 85, 86, 100, 102, 104, 113, 128, 138, 189, 190
stochastic model, 2
stratification, 176
stress, 23, 188, 190
structure, viii, 39, 42, 43, 46, 49, 50, 53, 57, 59, 62, 65, 68, 78, 83, 85, 86, 98, 117, 118, 121, 188, 195
supplier, 115

suppliers, 115
Switzerland, 191
symmetry, 190
synthesis, 154, 185, 212

T

Taiwan, 191
target, 16, 25, 26, 28, 35, 37
teams, 31
techniques, vii, viii, ix, x, 1, 11, 32, 39, 41, 52, 105, 120, 175, 176, 187, 188, 190, 215, 230
technologies, x, 187, 190
technology, 187, 190
teeth, 109, 118, 119, 130, 135, 143, 144
temperature, 188
test data, 176
testing, 8, 11, 12, 23, 120, 121, 125, 133, 142, 144, 179, 180, 181, 183, 185, 191
time pressure, 12
titanium, 190
tooth, 90, 113, 118, 119, 129, 130, 135, 137, 139, 143, 144
torsion, 47, 104, 107, 109
total product, 28
training, 13, 178, 179, 180, 181, 182, 228
trajectory, 195, 198, 200, 203, 204, 205, 206, 216, 221
transducer, 119, 120
transformation, 43, 71, 88, 194, 197
transformation matrix, 71
transmission, 101, 105, 106, 160
treatment, 5, 24, 79, 196

U

United Kingdom (UK), 148, 173, 189, 191, 192
updating, 41, 49, 52, 53, 54, 56, 58, 59, 65, 77, 78, 79, 142
USA, 117, 172, 192, 220

V

vacuum, 188
variables, 6, 60, 62, 66, 69, 71, 158, 179, 180, 211
variations, x, 109, 187, 190
vector, ix, 42, 59, 60, 62, 66, 67, 70, 79, 86, 88, 90, 93, 94, 97, 98, 156, 157, 161, 163, 175, 176, 178, 179, 180, 185, 198, 201, 202, 212, 216, 217

velocity, 40, 67, 69, 74, 75, 82, 94, 135, 158, 159, 160, 161, 176, 194, 196, 197, 198, 201, 230
versatility, 100
vibration, viii, 30, 39, 69, 82, 98, 104, 105, 107, 109, 110, 111, 112, 113, 114, 116, 118, 145, 188, 190, 201
vision, 192

W

Washington, 184

wear, vii, viii, 2, 3, 4, 7, 8, 11, 13, 15, 16, 17, 18, 19, 20, 21, 22, 23, 26, 28, 29, 30, 32, 36, 37, 117, 118, 119, 129, 130, 131, 132, 133, 138, 139, 141, 142, 143, 144, 147, 148, 212, 229
weight reduction, viii, 81, 102
wood, x, 193, 194, 195, 196, 200, 201, 202, 205, 206, 207, 208, 209, 210
workers, 15

Y

yield, 144